F V

 St. Louis Community College

Forest Park
Florissant Valley
Meramec

Instructional
St. Louis, M

D1116569

An OPUS book

THE PROBLEMS OF BIOLOGY

OPUS General Editors

Keith Thomas
Alan Ryan
Walter Bodmer

OPUS books provide concise, original, and authoritative introductions to a wide range of subjects in the humanities and sciences. They are written by experts for the general reader as well as for students.

The Problems of Science

This group of OPUS books describes the current state of key scientific subjects, with special emphasis on the questions now at the forefront of research.

The Problems of Biology

JOHN MAYNARD SMITH

OXFORD UNIVERSITY PRESS
New York   Oxford

Oxford University Press

Oxford   New York   Toronto
Delhi   Bombay   Calcutta   Madras   Karachi
Petaling Jaya   Singapore   Hong Kong   Tokyo
Nairobi   Dar es Salaam   Cape Town
Melbourne   Auckland

and associated companies in
Berlin   Ibadan

Published by Oxford University Press, Inc.,
200 Madison Avenue, New York, New York 10016

First published in 1986 as an Oxford University Press paperback and simultaneously in
a hardback edition, Walton Street, Oxford OX2 6DP

Oxford is a registered trademark of Oxford University Press

British Library Cataloguing in Publication Data
Smith, John Maynard
The problems of biology.—(OPUS)
1. Biology
I. Title   II. Series
574   QH307.2
ISBN 0-19-219213-2
ISBN 0-19-289198-7 Pbk

Library of Congress Cataloging in Publication Data
Maynard Smith, John, 1920–
The problems of biology.
(An OPUS book)
Bibliography: p.
Includes index.
1. Biology   2. Biology—Philosophy.
I. Title.   II. Series: OPUS
QH307.2.M38   1986   574   85-13763
ISBN 0-19-219213-2 (U.S.)
ISBN 0-19-289198-7 (U.S.: pbk.)

2   4   6   8   10   9   7   5   3   1

Printed in the United States of America

Preface

The suggestion that I write this book came from Henry Hardy of Oxford University Press. There were two reasons why I found the idea attractive. I was flattered by the thought that I might challenge comparison with Bertrand Russell's *The Problems of Philosophy*. More important, I am convinced, perhaps because I read the popular books of Eddington, Jeans, Einstein, Haldane, and Wells when I was at school, that the fundamental ideas of science can be explained to anyone willing to make the effort needed to understand them. I am also persuaded that to understand these ideas is one of the most exciting and rewarding experiences one can have.

I have tried to concentrate on problems rather than on accepted truths, but this at once raises a difficulty. It is impossible to discuss the questions we cannot answer without first answering those that we can. In biology, this means that we cannot think seriously about the major unsolved problems—in particular, the problems of development and of cognition—without first knowing something about molecular genetics and neurophysiology. The deepest knowledge we have acquired so far about living organisms is our knowledge of the molecular nature of heredity. It therefore seemed unavoidable that I explain this knowledge early in the book. There is a risk that this will prove hard going for those who do not know about DNA replication and protein synthesis, and boring for those who do. To the latter, I can only suggest that they skip rather lightly through Chapter 2; I can assure the former that to understand these ideas is worth the effort.

Much of biology, however, cannot be explained in molecular terms. In the face of this inability, there are two possible attitudes: I am not sure whether they should be seen as two world views, or merely as two different research strategies. One is to argue that the molecular approach has enabled us to understand heredity, and is already beginning to tell us things about how brains work and how organisms develop. What we should do, therefore, is to press on in the same way, making use of the wealth of new techniques of genetic manipulation that are now available. Even if we cannot yet see how such a problem as pattern formation in development will yield to genetic analysis, the experience of the past thirty years assures us that it will.

The second approach is to argue that wholes have properties that cannot be deduced from a knowledge of their parts. It follows, therefore, that development, behaviour, perception, and so on must be studied directly, and laws appropriate to those levels of organization be sought. It may later be possible to interpret those laws in molecular terms, but they cannot be deduced from molecular biology. After all, we do not expect to deduce the laws of economics from neurophysiology, even though we assume that they are the consequences of individual decisions to buy or sell, and that such decisions are the consequences of neural activities in people's brains.

I see myself as a neutral in this debate. Most problems are best solved by starting at both ends, and trying to meet in the middle, as I shall explain in more detail when discussing the problems of perception. I have no doubt that the outstanding problems in biology will require such a dual attack: in contemporary jargon, both top down and bottom up. But to be neutral in such controversies is to invite the hatred and contempt of both sides, so I want to say a few words to ward off such hostility.

First, a word to the molecular biologist. Members of that profession are often hostile to holistic ideas, and they are often justified. We must distinguish carefully between two kinds of holistic approach. The first, which seems to me entirely reasonable, is to assert that there are phenomena at higher levels which are likely to be discovered only by people who work at those levels, even though they are entirely consistent with known physics and chemistry, and may later be accounted for in those terms. A classic example is Mendel's discovery of the laws of heredity, which could not at that time (or ever?) have been deduced from molecular biology, but which have since been explained in those terms. The second, and illegitimate, type of holistic argument is to claim that, because we do not at present understand some phenomenon, there must be some vital force responsible for it. I can remember precisely this argument being used to persuade me that a chemical explanation of heredity was in principle impossible. Of course, there is no way we can be certain that there are not kinds of forces of which we are at present unaware, but our inability to understand something is not evidence of their existence. As it happens, I do not understand how modern sewing-machines work, but this does not lead me to suppose that the laws of topology have been broken: indeed, I feel confident I could find out if someone would let me take one to pieces. Molecular biologists are quite right to disbelieve in the *élan vital*.

Holists are, I think, in a weaker position, if only because in recent years progress has been so much faster from the bottom up than from the top down. Yet I do share their conviction that there are laws that can only be discovered by research on whole organisms, and on populations of organisms. Almost all my own work has been done at those levels. What should be the attitude of a biologist working on whole organisms to molecular biology? It is, I think, foolish to argue that we are discovering things that disprove molecular biology. It would be more sensible to say to molecular biologists that here are phenomena that they will one day have to interpret in their terms.

There is one important area of biology whose problems I have hardly touched on in this book: this is ecology. One of the important theoretical inputs to ecology is the idea of evolution by natural selection, and that I have of course discussed. But there are other ideas in ecology that have more in common with economics than with anything in contemporary biology, and these I have not dealt with. The reason is that these economic ideas are so different from those discussed in the rest of the book that I felt that any attempt to treat them would tend to confuse the picture.

I have tried, then, to present what seem to me the fundamental ideas of biology, and some of its major unsolved problems. Although, occasionally, I tell something of the history of a subject, I am not trying to contribute to the history of ideas; I do no more than use the past to illuminate the present. Also, I have made no attempt to give references for the statements I make. Detailed references would be of interest only to professionals, who would know most of them anyway. My reason for referring to some scientists by name is merely to show that science is something done by people, and not written on stone. My choice of names to mention is in part arbitrary: it is not an attempt judiciously to apportion credit. I have used notes in the main to introduce additional ideas that seem to me amusing, but which would interrupt the flow of the argument if included in the body of text.

I am grateful to Henry Hardy for provoking me into attempting the book, and prodding me into finishing it. My colleagues at the University of Sussex have provided the essential background of ideas and argument. Two of them must be mentioned by name. Paul Harvey read the whole manuscript, and made many helpful suggestions. To Brian Goodwin I have not shown a single word, for fear that he would make me rewrite the whole thing. However, those who know us will see that much of the book is a debate between us, in which he is not allowed to get a word in edgeways.

JOHN MAYNARD SMITH

Contents

List of figures

1

The definition of life

With the advent of space travel, the question has become more pressing: how shall we decide whether something is alive? If there are, on other worlds, objects as large and complex as elephants and oak trees, most of us are reasonably confident that we could recognize them as living. We are less confident about objects as small and, relatively, as simple as viruses and bacteria, but the difficulty arises because of the inadequacy of our sense organs rather than of our concepts. If we had eyes which magnified like an electron microscope, we would probably recognize a bacterium as alive.

Can we make more precise the criteria we would use? I suggest that there are two relevant properties:

i) Although the forms of living organisms remain constant, the atoms and molecules of which they are composed are constantly changing; in other words, they have a 'metabolism'.

ii) The parts of organisms have 'functions'; that is, the parts contribute to the survival and reproduction of the whole. Thus your legs are for walking with, your heart is for pumping blood round your body, and the feathery seed-heads of dandelions help to disperse the seed.

The major part of biology is concerned with these two properties of organisms: biochemistry and physiology with the former, genetics and evolution theory with the latter. In this chapter, I will try to make these concepts more precise, and ask how they are related and whether they are unique to living things.

Organisms are not the only objects which maintain a fixed form despite a continuous change in their constituent molecules. Consider two non-living examples. When you pull the plug out of a basin, the escaping water will form a rotating mass surrounding a narrow column of air; that is, it will form a vortex. If the level in the basin is kept constant by adding water, the form of the vortex remains constant, while the molecules of water pass through it and down the plughole. To take a second example, if you light a Bunsen burner, the flame has a fixed form, with different colours in different regions. Again, so long as there is a flow of gas, the form remains constant.

Both flame and vortex are examples of 'dissipative structures'. They are structures whose maintenance requires a continuous input of energy, and whose effect is to dissipate that energy. In the vortex, the energy is the potential energy of the water, which is dissipated as the water falls. In the flame, the energy from chemical reactions is dissipated as heat. As soon as the supply of energy stops, the form disappears. (In this respect, the conditions which preserve the structure are the exact opposite of those which preserve the form of a building or a snowflake, which last better in the absence of an input of energy.)

To what extent are living organisms dissipative structures? It is not true that all the atoms in an organism, as in a vortex or a flame, are in a state of flux. Many large molecules (proteins and nucleic acids) remain unchanged throughout the adult life of an organism, and are composed of the same atoms at the end of adult life as at the beginning. The extent to which this is true varies greatly between organisms (it is more true of adult insects than of vertebrates), between parts of organisms (it is more true of your brain than your intestine), and between kinds of molecules. In some quite complex organisms (e.g. some insects), this permanent molecular structure is so arranged that the animal can be freeze-dried, so that almost all chemical reactions cease, and can then be brought to life again.

Nevertheless, it is true that the maintenance of the living state requires a constant flow of energy through the system. A freeze-dried insect is not alive: it was alive, and may be alive again in the future. Energy must be supplied either in the form of suitable chemical compounds or as sunlight, and in either case atoms are continuously entering and leaving the structure of the organism. The difference is that the structure of even the simplest organisms is enormously more complex than that of a vortex or a flame, and that the flow of energy is controlled. This control depends on the continued existence of the large molecules mentioned in the last paragraph, and of larger structures made from them. The way these controls operate is the main subject-matter of biochemistry and physiology.

The use of the word 'control' brings us sharply up against the second characteristic of life: that the parts of living organisms contribute to the survival of the whole. Living things are not the only objects that have 'controls'. One of the simplest controls is the governor of a steam engine, whose rotating arms swing out further as the engine goes faster, and, by reducing the supply of steam, control the engine speed.

It was this parallel between machines and organisms which provided one of the central arguments of Natural Theology— that is, of that branch of theology which seeks to give reasons for a belief in God derived from a consideration of the natural world. William Paley, the most famous of the Victorian natural theologians, argued as follows. If we look at a piece of machinery such as a watch, it is apparent that its structure has a purpose— telling the time. From this we deduce that the watch must have been designed, and hence that it had a designer. Since we see that living organisms have the same characteristics—that their parts have functions or purposes—we can again deduce that they were designed. Hence we can deduce the existence of God, the designer.

Before discussing how modern biologists interpret functions, one difference between the views of theologians and biologists needs comment. We agree that the heart has a function: we differ only as to how it acquired that function. But do elephants or oak trees have functions? My own view is that they do not: it is as senseless to ask what an elephant is for as it is to ask what an electron is for. But for some theologians, the world was created for men to live in, and so animals have a purpose: to serve man. To the question 'what are seals for?' one theologian answered that they are there to give polar bears something to eat, because otherwise the bears would come south and eat us. Apart from the infinite regress implicit in this reply ('What are polar bears for?' and so on), few biologists would accept the anthropocentric assumption. The relevance of this distinction, between the parts of organisms having functions, and whole organisms or populations of organisms having them, will emerge in Chapter 5.

Yet I agree with Paley that the heart has a function, even if we disagree about the elephant. But biologists no longer accept the argument from design, because they have an alternative explanation of how the parts of organisms come to have functions. This alternative is Darwin's theory of evolution by natural selection.

Darwin's theory can be formulated as follows. Given a population of entities having the properties of multiplication, variation, and heredity, and given that some of the variation affects the success of these entities in surviving and multiplying, then that population will evolve: that is, the nature of its constituent entities will change in time. Of the three essential properties, multiplication means that one entity can give rise to two, and variation that not all entities are identical. Heredity means that like begets like. The process is illustrated in Figure 1; the essential point is that when multiplication occurs, kind A usually gives rise to A, kind B to B, and so on. If heredity were exact, evolutionary change would eventually slow down and stop; continuing evolution requires that heredity is inexact, so that new variants arise from time to time.

Given a population of this kind, we can say rather more than that evolution will occur; we can say something about the kinds of changes which will occur. The population will come to consist of entities with properties which help them to survive and

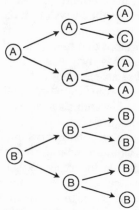

Figure 1. Heredity and variation.
The meaning of heredity is that, when multiplication occurs, like gives rise to like: A gives rise to A, and B to B. Variation requires that this rule is occasionally broken, as when A gives rise to C.

reproduce. Note that this does not give us a fully predictive theory of evolution, because the course of evolution will also depend on the repertoire of variation which arises. But we do expect organisms which have evolved by natural selection to have organs which help them to survive and reproduce: that is, we expect them to have parts with 'functions'.

So formulated, the theory of evolution is not a falsifiable scientific theory in the sense required by Karl Popper, who demands of a theory that it should rule out certain kinds of events which, if they do in fact occur, can be used to show that the theory is false. As formulated in the last two paragraphs. Darwin's theory is not falsifiable. It is a statement of a logical deduction: if certain things are true, then evolution follows. A scientific theory must say something about the world, and not merely about logical necessity. Darwinism as a testable scientific theory can take various forms. I will give it first in the form in which Darwin himself proposed it, and then in the 'neo-Darwinist' form in which most biologists hold it today.

For Darwin, the theory had two components. The first was the proposition that all living things on earth are descended, with modification, from one or a few simple original forms. He did not claim to account for the origin of those first living things. This part of Darwin's theory, sometimes referred to as the 'fact' of evolution, is almost universally accepted by biologists today. The second component of the theory was that the major (but not the only) cause of evolutionary change was natural seletion. As a subsidiary cause of evolution, Darwin accepted 'the effects of use and disuse'; to over-simplify, he thought that organisms acquired during their lifetimes changed characteristics (called by biologists 'characters' or 'traits'), which they would then pass on to their offspring.

The claim of descent with modification is clearly falsifiable; as J. B. S. Haldane remarked, a single fossil rabbit in Cambrian rocks would be sufficient, because the first fossil mammals are found in rocks some 400 million years later than the Cambrian. The theory that natural selection is one cause of evolution is not so easily falsifiable. It would be falsified if one could show that organisms lack one of the three necessary properties of multiplication, variation, and heredity, but I know of no one who has ever attempted to do so.

Thus I think it may be an error to try to force Darwin too strictly into the Popperian mould; after all, Popper's philosophy was derived from a study of physics. Darwin's claim that evolution has happened certainly falls within Popper's criterion. His proposed mechanism of natural selection, however, does not, because he put it forward as only one among several possible mechanisms and not as the only one, so that any observation which could not be explained by selection could always be explained in some other way. As I shall explain later, modern evolution theory is closer to meeting Popper's criterion, because it is no longer so ready to accept alternative explanations of adaptation.

Darwin's position is best summarized as follows. He started with the observation that organisms are adapted to survive in particular environments, and that they have parts which function so as to ensure that survival. That is, his starting-point was that any adequate theory of evolution must explain adaptation. In his theory of natural selection, he pointed out that organisms do in fact have the properties of multiplication, variation, and heredity, and that it is a necessary consequence of this that they should become adapted for survival. That is, he showed that there is a necessary connection between one set of observations, concerning reproduction, and another set, concerning adaptation.

Our main difference from Darwin lies in our possession of a theory of heredity, outlined in the next chapter. This has led us to develop Darwin's ideas in two main ways:

i) As discussed in detail later, we no longer think that changes that occur during an individual lifetime as a consequence of the use or disuse of organs alter the nature of the children produced by that individual. Hence we no longer regard the 'inheritance of acquired characters' as a significant cause of evolution. In this sense, we are more Darwinian than Darwin, since we regard natural selection as the only relevant mechanism leading to the evolution of adaptation, instead of merely the most important one.

ii) We are more interested in other possible causes of evolutionary change. There are various possibilities. First, there are changes (variously referred to as 'genetic drift' and as 'non-Darwinian evolution') which occur because the traits concerned have so little effect on survival and reproduction that the changes are essentially random. Second, the entities which are subject to evolution by natural selection may not be individual organisms, but either larger entities (populations) or smaller ones (genes, or groups of genes). Third, changes may occur because information can be transmitted between generations not only genetically, but also culturally.

These are changes of emphasis rather than substance. The central point remains that Darwin provided a theory which predicts that organisms should have parts adapted to ensure their survival and reproduction. This has led to the suggestion that life should be defined by the possession of those properties which are needed to ensure evolution by natural selection. That is, entities with the properties of multiplication, variation, and heredity are alive, and entities lacking one or more of those properties are not.

There is much to be said for this definition. Consider, for example, the question of whether fire is alive. One answer might be that it is not, because although it has a structure which remains constant despite a continuous change of substance, that structure is too simple to qualify as living. But that leaves us with the unilluminating question of how complicated an object must be to qualify as living. It is more illuminating to say that fire is not alive because, although it has the properties of multiplication (one fire can light another) and variation, it lacks heredity. Thus, although fires vary in size, colour, and temperature, their nature at any instant depends on immediate circumstances—the supply of fuel, how much wind there is, and so on—and not at all on whether they were started with a match or a cigarette-lighter. Because, with fires, like does not beget like, fires cannot evolve by natural selection, and so do not become complex, and do not acquire organs to keep them going.

From this preliminary survey, two distinct pictures of living organisms have emerged. One is of a population of entities which, because they possess a hereditary mechanism, will evolve adaptations for survival. The other is of a complex structure which is maintained by the energy flowing through it. The first view is pursued in Chapters 2 to 5, and the second in Chapter 6. Perhaps the hardest part of biology is to see how the two pictures fit together. Most questions in biology require at least two answers, one appropriate to each of the two pictures. Suppose we ask, 'why does the heart beat?' One answer will be concerned with how differences in heart shape are inherited, and how they influence the survival of their possessors, and will lead to the conclusion that the function of the heartbeat is to pump blood round the body. The other will be concerned with the rhythmical properties of heart muscle, and with the nerves supplying the heart, and will lead to an account of heart physiology. Rather unhappily, these two kinds of answers have been called, respectively, the 'functional' and 'causal' explanations. The words are unhappy, because the 'functional' explanation is also a causal one, but on a longer time-scale. It is perhaps better to refer to the two kinds of explanation as the ultimate and the proximal causes of the heart beating. The important thing to understand, however, is that the two explanations are not alternative and mutually exclusive; both can be true, and they can illuminate one another.

2

Heredity

I have argued that the ability of like to beget like is the most fundamental characteristic of life. It is also one of the best understood. Although this book is a discussion of the problems of biology, the first half of this chapter outlines the theory of genetics, which is perhaps the best established and least problematic area of biology. My excuse is that one can only understand the unknown in terms of the known. For those unfamiliar with molecular genetics, this chapter may prove difficult, but the ideas it contains do constitute the core of modern biology, so it is worth making an effort to master them.

There have been four main stages in the growth of our knowledge of genetics: Weismann's concept of the independence of germ line and soma; the establishment of an atomic theory of heredity following the rediscovery of Mendel's laws in 1900: the chromosome theory of heredity, based in particular on the work of T. H. Morgan and his colleagues on the fruitfly *Drosophila*; and the growth of molecular genetics, initiated by the discovery of the structure of DNA in 1953 by Watson and Crick. In fact, Mendel's work was carried out before Weismann's, but it did not become part of the corpus of biology until later, so Weismann's ideas will be discussed first.

The German biologist August Weismann, writing at the end of the last century, argued that two independent processes of cell division start from the fertilized egg. One leads to the adult body, or 'soma', and the other, constituting the 'germ line', leads to the germ cells—egg and sperm—which are the starting-point of the next generation. The soma is mortal, but the germ line potentially immortal. Before Weismann, most evolutionary theorists, and in particular the Frenchman Jean Baptiste Lamarck (1744–1829), had supposed that the main cause of evolutionary change was the 'inheritance of acquired characters'. Darwin, too, as we have seen, accepted this possibility, under the description 'the effects of use and disuse'. However, if Weismann was right, there was no way in which the characteristics acquired by an adult—the blacksmith's large muscles are the classic example—could alter the nature of the offspring.

In support of his view, Weismann pointed out that in most animals the cells which are going to give rise to germ cells are set aside very early in development. If they are destroyed, they cannot be replaced, and the animal will be sterile. This does not prove Weismann's case, for two reasons. First, no early separation of the germ line occurs in plants, yet today we are as ready to reject the inheritance of acquired characters in plants as in animals. More fundamental, all the energy and material needed for the growth of the germ cells is supplied by the rest of the body, so there are plenty of opportunities for the body to influence the germ cells.

Weismann did offer some direct experimental evidence for the non-inheritance of acquired characters, and a great deal more is available today. However, I think that his main reason for rejecting the possibility is that he could not imagine a process whereby it could happen. He was one of the first people to see that what matters in heredity is a flow not of matter or energy but of information. Suppose a blacksmith does develop big muscles. How could that so alter the sperm he produces that his sons would resemble him? The sperm cell has no muscles. In modern terms, the big muscles of the father would have to be translated into some coded form, which would later, in his sons, be translated back into big muscles again.

Today, most of us own machines which translate information from one form to another. For example, a tape recorder can translate sound into patterns of magnetization on a tape, or vice versa. That Weismann understood the significance of the information analogy is shown by his remark that to accept the inheritance of an acquired character would be 'very like supposing that an English telegram to China is there received in the Chinese language'.

Although Weismann's ideas are fundamental for an understanding of evolution, modern genetics originated with the rediscovery of Mendel's laws in 1900. Gregor Mendel, a contemporary of Darwin's, and the head of a monastery in Moravia, had been trained in the physical sciences, and this training shows up in his approach to heredity. Perhaps the most interesting thing about him was not the conclusion he reached, but the way he reached it. The first critical step was to ask the right question. The obvious question to ask would be: why do mice give birth to mice and elephants to elephants? Although obvious, this would not have been a fruitful question to ask, because there was no way in which it could be answered. Instead, Mendel was concerned, as most geneticists since, with the causes of the differences between fairly similar organisms—between brown and white mice, or, in Mendel's case, between tall and short peas. He also had the good judgement to concentrate on those cases in which his pea plants could be classified into one of two distinct classes, with no intermediates.

By recording the numbers of plants belonging to different categories in successive generations, he was led to formulate an atomic theory of heredity. In effect, he argued that the cause of the difference between tall and short, or smooth and wrinkled, peas was the presence of hereditary factors, which we now call genes.[1] For each trait (for example, the trait of being tall or short, or of being smooth or wrinkled), each individual receives two genes, one inherited from each parent: thus an individual might receive two genes for tallness, or two for shortness, or one of each. When the individual produces a germ cell (egg or sperm), one of the two genes, chosen randomly, is transmitted to that cell, and hence to the offspring.

Mendel was led to this conclusion by the fact that individuals with different characteristics appeared in families in fixed ratios. For example, if a first-generation hybrid between a brown and an albino mouse is crossed back to an albino, the offspring will be in the ratio 1 brown to 1 albino: if two hybrids are mated to one another, the ratio will be 3 brown to 1 albino. There is an uncanny resemblance between this reasoning and that which had earlier led John Dalton to an atomic theory of chemistry. Dalton never saw an atom, or the effects of a single atom. He reasoned that they must exist because of the fixed and numerically simple proportions in which elements combine to form compounds. Unlike Dalton, Mendel was not appreciated until after his death. One reason for this may have lain in the unwillingness of biologists to accept the highly abstract nature of his theory. By this I do not mean that he did no experiments, but that he explained his results by hypothesizing the existence of entities for which he had no direct evidence. In general, biologists have been distrustful of such leaps of the imagination, although later geneticists have followed Mendel's example: a recent case is the postulation by Jacob and Monod of 'repressors', described on p. 65. The difference is that, with the invention of new techniques, the gap between postulating an entity and identifying it physically is much shorter than it was in Mendel's case.

In the thirty years between the publication of Mendel's results and their rediscovery, great progress had been made in describing the structure and behaviour of cells. In particular, it was known that each cell contains a nucleus which can be seen to contain a number of threads, the chromosomes. Further, the chromosomes had precisely the properties postulated by Mendel as belonging to his hereditary factors. That is, each body cell contains two sets of chromosomes (for example, 23 pairs in human cells), but the gametes (egg in the female and sperm in the male) receive only one of each pair. The chromosomes themselves cannot be the Mendelian factors, or genes, because far more than 23 kinds of gene are required to account for human differences. But the chromosomes could be carriers of the genes, and, since chromosomes are threads, it was reasonable to guess that the genes are arranged along the chromosomes, like beads on a string.

That this is indeed the case, or roughly the case, had been established by the end of the 1930s, primarily by the American T. H. Morgan and his colleagues Calvin Bridges, A. H. Sturtevant, and H. J. Muller, working on the fruitfly *Drosophila melanogaster*. In essence, their procedure was to follow simultaneously the numbers of individuals with various traits (white eyes, short wings, bent bristles, etc.) in successive generations, and the structure and behaviour of the chromosomes. Usually, Mendel's laws were obeyed, but they found many cases in which they were not. The commonest departure arose when two traits were determined by genes on the same chromosome (that is, were genetically 'linked'), so that the traits did not segregate independently of one another, as Mendel claimed that they should. A number of rarer departures from Mendel's laws were found, and in each case they were able to find a corresponding abnormality in the chromosomes, of precisely the kind required to explain the genetic findings.

The theory of heredity as it existed in the 1940s, then, was one of genes linearly arranged along chromosomes, influencing the characteristics of the organisms in which they found themselves, and being transmitted via egg and sperm to the next generation. There were, of course, complications. It was known that one gene could influence several traits (pleiotropism) and that a single trait could be influenced by many genes (polygenic inheritance). Major problems remained. How did genes influence the characteristics of organisms? How could they be transmitted unchanged to future generations?

The next decisive step was the discovery of the chemical nature of genes. To follow this, an irreducible minimum of biochemical knowledge is necessary. First, we must distinguish two classes of molecule found in cells—'metabolites' and 'macromolecules'. A metabolite is a small 'organic' (that is, carbon-containing) molecule, with up to about 50 atoms. Typical examples are sugars like glucose, and amino acids like leucine. A macromolecule is a long string, or polymer, formed of a series of similar building blocks; these building blocks are themselves metabolites. For our present purpose, two kinds of macromolecule are important, proteins and nucleic acids. Proteins are strings of 20 kinds of amino acids, which usually fold up to form a large globular molecule. They compose much of the actual substance of the body (muscles, tendons, the lens of the eye, for example), and are the organic catalysts responsible for the chemical reactions occurring in the body, as described in Chapter 6. Nucleic acids are strings of four kinds of 'nucleotides': each nucleotide consists of a common piece (which joins to other identical pieces to form the 'string'), and, sticking out sideways, one of four different 'bases' (adenine, thymine, guanine, or cytosine).

One kind of nucleic acid, DNA or desoxyribosenucleic acid, was known to be present in chromosomes, and hardly anywhere else. It would, therefore, have been natural to suppose that genes were made of DNA. However, people were rather slow to draw this conclusion, for two reasons. One was that an erroneous idea existed as to the structure of DNA. According to the 'tetra-nucleotide' theory, there was a central backbone, surrounded by rings of four nucleotides, each ring containing one nucleotide of each kind. If this was true, any piece of DNA was just like any other piece, whereas every gene must be different from every other one. A second reason for doubting that genes were made of

DNA was that proteins are chemically much more interesting, and it was already known that at least some genetic abnormalities are caused by the absence of specific proteins. One way of making sense of this was to suppose that genes were made of proteins, and that the DNA merely provided the string on which the protein beads were strung.

However, evidence was accumulating from the study of microorganisms that DNA might have some genetic effects, and this forced a reconsideration of its chemical structure. The answer is the now famous double helix discovered by James Watson and Francis Crick in Cambridge in 1953. The critical features are shown in Figure 2. DNA consists of two strands, each with a backbone carrying the bases sticking out sideways. On any given strand, the bases can be in any order, but there are rules governing the pairing between the bases on the one strand and those on the

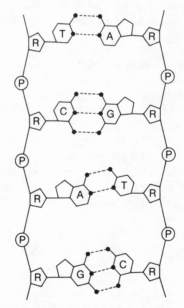

Figure 2. The structure of DNA.

The letters A, T, C, and G represent the four bases adenine, thymine, cytosine, and guanine, respectively; R and P represent the ribose sugar and the phosphate forming the backbone of each strand. The dotted lines represent the bonds linking complementary bases.

other; adenine always pairs with thymine, and guanine with cytosine. When the molecule replicates (Figure 3), the two strands come unzipped, and a new strand is synthesized against each pre-existing one. Provided that the above rules of base pairing are guaranteed by the laws of chemistry, it is easy to see how replication gives rise to two daughter molecules identical to the original one. The problem of how this specificity of pairing is guaranteed is discussed on p. 117. For the present, what is important is that the Watson–Crick structure does explain how DNA can carry information, and how it can replicate. It can replicate because the specificity of base pairing ensures that the daughter molecules are identical to the original one. It can carry information because each DNA molecule can have a different

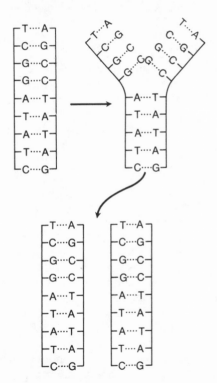

Figure 3. The replication of DNA.

sequence of bases, just as every paragraph in this book has a different sequence of letters. There remains the problem of how DNA ever gets round to doing anything; that is, of how the information carried in the base sequence is made use of.

Genes, then, are simply lengths of DNA. But what do they do? There is a short and simple answer to this question. Genes specify the kinds of proteins a cell can make. That short answer gives rise to two further questions. How do genes specify proteins? How does the ability of genes to specify proteins enable them to control development? The second of these questions is discussed in Chapter 9; the first is discussed now.

DNA consists of a string of four kinds of bases; protein of a string of 20 kinds of amino acid. These two sequences are connected by the 'genetic code', which is in effect a listing of which groups of bases specify which amino acids. I will describe in a moment how the decoding is carried out, but first, a word about what kind of code it is. The code is simple: each group of three bases specifies an amino acid. Clearly, groups of two bases would be insufficient, because there would be only $4^2 = 16$ ordered pairs (that is, there are four ways of choosing the first base, and, for each choice, four ways of choosing the second), and hence only 16 different amino acids could be specified. Groups of three bases provide $4^3 = 64$ ordered triplets, which seems excessive to code for 20 amino acids. The code in fact incorporates a degree of redundancy, so that in some cases as many as six different triplets code for the same amino acid. Some triplets, too, are translated as 'stop reading here', or 'end of protein'.

The previous paragraph assumes the existence of a decoding device in the cell, able to translate particular triplets into particular amino acids: such a decoding device is analogous to a machine which could receive a message in Morse code (corresponding to the base sequence in DNA) and produce a version written in the roman alphabet (corresponding to the amino acid sequence of the protein). The nature of this decoding device is shown in Figure 4; it is complicated, but unfortunately one must understand the complications in order to understand some of the fundamental questions that can be asked about the nature, origin, and evolution of the code. It is important to appreciate that the processes which are illustrated in Figure 4 are going on on the surface of a special structure (it is called a 'ribosome'), which acts

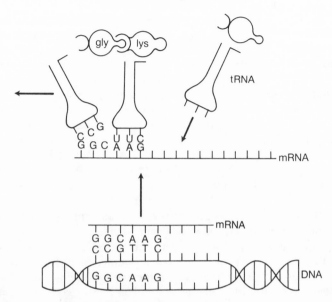

Figure 4. Protein synthesis.

Below is shown a DNA molecule, with an mRNA copy being made of one of its strands. Above, the same mRNA molecule is shown after it has moved to the cytoplasm, where tRNA molecules are pairing with it. The first tRNA, with the anticodon CCG, has already released its amino acid, glycine, and is now leaving the site of synthesis. The second tRNA, with the anticodon UUC, is paired to the codon AAG in the message, and its amino acid, lysine, is being attached to the glycine. A third amino acid is being brought by a third tRNA. Note that tRNA molecules arrive at random, but only those which happen by chance to have the right anticodon will remain and contribute their amino acids to the growing protein chain. Note also that the base A pairs both with T, in DNA, and with U, in RNA.

rather like the assembly line on which cars are put together, holding the various molecules in position.

First, we must introduce two additional kinds of molecule. Both are made of RNA, a nucleic acid which differs from DNA in being single-stranded, and in having the base thymine replaced by the very similar uridine. The first type is 'messenger' RNA (mRNA for short), which is made in the nucleus as a copy of a DNA gene, and which then passes into the cytoplasm (the part of the cell outside the nucleus) to the ribosome assembly line. The second type is 'transfer' RNA (tRNA), which brings amino acids to the assembly line. The essential steps in decoding are then as follows:

i) A particular gene is 'transcribed' into mRNA; that is, a complementary RNA strand is synthesized by base-pairing with one of the two DNA strands. As shown in the figure, the DNA has to be partially unzipped to allow this. Note that only one of the two DNA strands is transcribed; the other is there just to facilitate replication, as was illustrated in Figure 3. The transcribed strand is the one that carries the genetic information. This message then leaves the nucleus, and passes to the ribosome assembly line.

ii) Suppose that, as shown, the first triplet of bases in the message is GGC; this is the first 'codon'. There will be, in the cytoplasm, tRNA molecules carrying the complementary triplet, or 'anticodon', CCG. Because of the chemical affinity of G for C (and of C for G), one of these molecules will pair, as shown in the figure, with the first codon on the message.

iii) The tRNA molecules with the anticodon CCG also have attached to them the amino acid glycine. In this way glycine becomes the first amino acid in the protein, being carried into position on the ribosome assembly line by its tRNA. Once this has happened, the glycine molecule is separated from its tRNA, which drifts away: the glycine is held in position on the ribosome.

iv) The next triplet of bases in the message (AAG in the figure) then pairs with a tRNA carrying the anticodon UUC, and the amino acid lysine; thus the second amino acid of the protein, namely lysine, is lined up with the first. The two amino acids are then joined together.

v) For each successive triplet in the message, a new amino acid is added to the protein in the same way.

There is a 'chicken and egg' character to this decoding machinery. If it is to work properly, there must be a number of special protein and RNA molecules present, but these molecules are themselves the products of an earlier decoding process. This need not worry us until we come to discuss the origin of life in the last chapter: in existing organisms, proteins never arise except from a decoding process, and decoding never happens without proteins. For the present, what is important is that the nature of the code (for example, the fact that GGC codes for glycine) depends on the existence of tRNA molecules with a CCG anticodon at one end and a glycine molecule attached at the other. How does this come about? The tRNA molecules are themselves copies of DNA genes in the nucleus. When they are first made, they do not have a glycine molecule attached. That attachment is performed in the cytoplasm by a special protein, which is able to recognize both the glycine molecule, and the end of the tRNA molecule, and stick them together. Proteins acting in this way are called 'enzymes'; their method of working is described on p. 61. Of course, these enzymes are also coded for by a DNA gene in the nucleus.

Thus there are in the cytoplasm many different kinds of tRNA molecule, each kind being characterized by two 'labels': at one end is the anticodon (which pairs with a codon in the message), and at the other end an attachment site for an amino acid (which is recognized by an enzyme carrying that particular amino acid). The nature of the code (that is, for example, that GGC codes for glycine) depends entirely on which anticodon and which attachment site occur together on a tRNA molecule. It follows that there are some kinds of changes in the nuclear DNA (such changes are called 'mutations'; see p. 22) which will cause a change in the code. This could be caused either by a genetic change which altered the relevant attaching enzymes, or by a genetic change which altered the anticodon of a tRNA molecule without altering the attachment site for the amino acid (or vice versa); such mutations do occur as rare variants.

Nevertheless, the code is universal: that is, GGC is translated as glycine in all organisms from viruses to man, and similarly for the other triplets. This suggests that either there is some 'best' code on which all organisms have converged, or that all existing organisms are descended from the same simple ancestor, and that the code has not changed since. The first alternative is unlikely. It is hard to see why a code in which GGC means glycine and AAG means lysine is either better or worse than one in which the meanings are reversed. Similarly, a language in which 'horse' meant a quadruped with cloven hoofs and horns, and 'cow' one with a single toe and a mane, would be neither better nor worse than English. Like human languages, the genetic code is to a considerable extent arbitrary.

It is easier to see why, once established, the code does not evolve. Thus suppose a mutation occurs in a complex organism which causes GGC to be translated as proline instead of the usual glycine. It is possible that in some protein it would be an improvement to change one particular glycine to proline. But glycine occurs in thousands of different proteins, and it is inconceivable that the change should be beneficial in most of them. By analogy, it may sometimes happen that a message is improved by changing, say, an s into a G: my secretary once, misreading my writing, replaced 'Sod's law' by 'God's law', and the sense was if anything improved. But to change all the s's in a book into G's would cause chaos.

Once the code is established, then, it is unlikely to change. Its universality is to be expected if all existing organisms have a single ancestor. Whether this means that life originated just once, or that it originated many times, each origin acquiring a different code, but that one origin gave rise to more successful competitors, we do not know.

One feature of the genetic decoding mechanism is that it appears to be irreversible. This was expressed by Crick as the 'central dogma' of molecular biology, that information can pass from DNA to protein, but not from protein to DNA. Before spelling out exactly what this means, it is worth asking whether translating machinery is necessarily irreversible. The example of a tape recorder shows that it is not: the same machine can translate information in the form of sound waves in the air into patterns of magnetization on a tape, and vice versa. However, translating machines can be irreversible. A record-player will translate information carried as undulations in a groove into sound, but not back again: you cannot make a record by shouting at the loudspeaker of a record-player. The central dogma in effect claims that the genetic system resembles a record-player, not a tape recorder. That is, if you introduce into a cell a protein molecule with a novel sequence of amino acids, the cell cannot synthesize nucleic acid molecules with a base sequence coding for that protein.

The central dogma is of course a theory, but there is no evidence to suggest that it is wrong. The irreversible step is that between RNA and protein. Although, typically, information passes from DNA to RNA, it can travel in the opposite direction. This is sometimes quoted as a contradiction of the central dogma, but that is a misunderstanding.

It has often been said—indeed I have said it myself—that the importance of the central dogma is that it provides a molecular explanation for Weismann's theory of the independence of germ line and soma. In a sense this is true. If the effect of use and disuse is to alter the nature of the proteins in the body, and if the replicable information passed on to future generations is carried by DNA, then Weismann has to be right if the central dogma is right. However, he would be right most of the time even if the central dogma is wrong, and protein sequences could be translated into DNA sequences, because most 'acquired characters' would not involve the synthesis of new kinds of proteins. When a blacksmith grows big muscles, he makes more of certain kinds of proteins, but it is doubtful whether he makes any new ones. If Weismann is wrong, I think it will be because there are ways of transmitting information between generations other than via DNA. I will return to this possibility at the end of this chapter. First, I want to comment on some features of the hereditary mechanism which may prove to be characteristic of all living things, whether on this planet or elsewhere in the universe. This will remain an untestable speculation unless and until we meet other forms of life, but speculation can be useful.

I can suggest three such general features: heredity is digital, it involves a phenotype–genotype distinction, and it makes possible the amplification of quantum events into macroscopic ones. I will explain these points in turn.

If an information system is digital, its symbols belong to one of a finite number of discrete categories, and their meanings depend on which category they are in. If, as may well be the case, there is variation within one of these categories, that variation is without meaning. It seems that all reasonably efficient communication systems are digital in this sense. The reason is that, with a digital system, small variations do not corrupt the message. For example, Englishmen pronounce the A in CAT in slightly different ways, but, provided the variation is not too great, there is no confusion between CAT and COT. If meaning depended on the value of a continuous variable, then every time the message was replicated it would change, however slightly.

In classical genetics, the 'phenotype' of an individual is its structure and behaviour, and its 'genotype' is its genetic constitution. The distinction reflects the more fundamental one between a mortal body and a potentially immortal genetic message. A man inherits his father's nose in a different sense from that in which he inherits his watch. In the latter case, the physical object is passed on; in the former, all that is passed on are genes that influence the shape of the nose as it develops. Strictly, not even the genes are passed on. The actual DNA molecules I received from my parents are not passed to my children: replicas of them are passed on.

Even if life elsewhere uses a genetic material different from DNA, I think that the distinction between genotype and phenotype will remain. There are two reasons for this. One is that most acquired characters are disadvantageous—they are results of injury, disease, and old age. Hence a hereditary mechanism which transmitted them would result in continued degeneration. A sharp distinction between soma and germ line makes it possible to prevent acquired characters from being transmitted. But there is a second reason. Bodies are selected to have properties ensuring growth and survival; these properties are likely to be incompatible with accurate replication. In the organisms we know, replication depends on an accurate fit between the surfaces of complementary molecules—that is, between DNA molecules. Such molecules do not have the active catalytic properties needed for metabolism, growth, and movement: these properties are provided by proteins. The distinction between genotype and phenotype reflects a division of labour between nucleic acids and proteins.

A third property which may be universal is that a change of a single molecule can, regularly and predictably, cause large-scale changes in the body.[2] I have so far said little about variation, the third property required for evolution along with multiplication and heredity. This is because variation is the obverse of heredity: it is what happens when the mechanism goes wrong. Heritable changes—called 'mutations'—occur if the genetic message is changed. Mutations occur mainly, but not exclusively, when the DNA is replicating; their frequency is increased by agents (in particular, highly reactive chemical substances) which interfere with the replication process. The results can range from the substitution of one base by another to larger rearrangements of the genetic material. But even a single base change in the germ line DNA can cause gross visible effects, because the changed DNA molecule is replicated, and its effects are further amplified during the decoding process. In this respect, living systems differ from non-living ones: you cannot make a visible change in the form or behaviour of a wave in the sea by moving one molecule, or even a million molecules. However, living systems do resemble machines and other human artefacts. You can alter the behaviour of a power plant by throwing a switch; it is true that to move a switch is to move more than one molecule, but there is nevertheless an amplification of a small signal into a large effect. This amplification becomes possible once a system is controlled, and is most dramatic in living things. The alteration of a single molecule in the egg of a fruitfly can cause the new individual to have an extra pair of wings.

This property of amplification makes it impossible to make long-term predictions about evolution, as one can in astronomy. It is sometimes said that mutation is 'random'. At a chemical level this is untrue, because different chemical agents cause different kinds of change. What is true is that the effects of a mutation on the form of the resulting organism cannot be predicted from a knowledge of the agent that produced it. The combination of the unpredictability of mutation, and the amplification of its effects, makes long-term predictions impossible. This does not mean that there are no generalizations that can be made about evolution, but it does make biology a different kind of science from physics.

It is now possible to summarize the neo-Darwinian theory of evolution. Organisms differ, partly because of the varying impact of the environment in their lifetime ('nurture'), and partly because of differences between the genes present in the fertilized eggs from which they developed ('nature'). Only the latter differences will be inherited. Evolutionary changes occur because some heritable differences affect the probability of survival and reproduction, so that the proportions of different genotypes in the population change generation by generation.

Few biologists doubt that this is part of the truth, but is it the whole truth? One qualification is that some evolutionary changes occur by chance, without natural selection. Another concerns the way in which selection acts, and in particular the target of selection: should we think of genes, or organisms, or populations, as being the entities which survive and reproduce, and upon which selection acts? These topics are discussed in Chapter 5. A more fundamental qualification concerns the nature of heredity.

Is it really true that nucleic acids are the only carriers of genetic information? The only serious alternative to natural selection as the cause of adaptation is Lamarck's notion of the inheritance of acquired characters. If nucleic acids are the only carriers of heredity, and if the central dogma is correct, then Lamarckism is ruled out. It seems unlikely that the central dogma will prove mistaken. But what of alternative carriers of heredity?

First, there is one argument which is commonly used in this context, but which seems to me mistaken. It is pointed out, quite correctly, that an egg which consisted of a bag of water containing all the necessary genes would not develop: other structures and substances are needed. But this is not evidence for an additional hereditary mechanism. Heredity requires that different types should exist, say A's and B's, and that like should beget like— A's should give rise to A's and B's to B's. To demonstrate non-DNA inheritance, one must show that two fertilized eggs differ in some respect other than their nucleic acids, and that when these eggs develop into adults which in turn produce eggs, those eggs differ in the same respect. That is rarely, if ever, the case, but there are phenomena which approach it, and which need some thought.

The cells of the body are not all alike: there are liver cells, kidney cells, and so on. Even if these cells are removed from the body and kept in tissue culture, they retain their characteristics through many cell divisions. There is good evidence that the differences between them are not caused by differences between the DNA sequence of their genes, although it may be that different genes are active in different kinds of cells. The nature of the cellular changes responsible for tissue differentiation are not well understood, but at the cellular level they do qualify as hereditary changes: each kind of cell gives rise to its own kind at cell division. However, it seems that when an egg is produced, it carries no memory of any differentiation that may have occurred in the organism that produced it. Occasionally, the restoration may not be complete. In organisms which reproduce asexually, without producing eggs, quite long-lasting changes can arise, without apparently requiring a change in DNA sequence. Perhaps the most striking example is the long-continued transmission of abnormal patterns of cilia (microscopic hair-like protrusions) on the surface of ciliated protozoans, demonstrated by the American geneticist Tracy Sonneborn and his students. My own view is that these exceptions are so infrequent that they are not a reason for abandoning the neo-Darwinist mechanism as the one which underlies the vast majority of adaptive evolutionary changes. But it is important that we should not forget them.

3

Sex, recombination, and the levels of life

This chapter looks at the relationship between genes and organisms in a new way. Figure 1 (p. 4) pictures the simplest possible population of replicating entities. If the world was really like that, then each individual today would have only one parent, and only one ancestor a hundred (or a million) generations ago. All the genes in that individual would be copies, with or without modification, of genes in a single ancestor a million generations ago. But the world is not like that. The genes in an organism today have come from many different ancestors. The most familiar reason for this is that most reproduction is sexual: an individual has two parents, not one. As we shall see, there are other and less familiar ways in which genes from different ancestors can come together in a single descendant.

This shuffling around of genes greatly complicates the analysis of evolution. Thus imagine that the living world was really as imagined in Figure 1. Then the criteria which would ensure the spread of some trait—say sharp teeth—in a population would be identical to those ensuring the spread of a gene causing that trait: if sharp teeth increase 'fitness' (that is, chances of survival and reproduction), then genes causing teeth to be sharp will increase in frequency. Further, since sets of genes stay together generation after generation, there is no 'conflict of interest' between the genes in an organism. But once we allow for the shuffling of genes, there is a whole new set of possibilities. As we shall see, a gene may increase in frequency not because it increases the fitness of its carrier but because it associates with other genes that do, or even because it is in a sense parasitic on other genes. Faced with these possibilities, we have to think very carefully about the ways in which natural selection acts. I start by discussing the more familiar phenomena of sex and recombination.

Although, as human beings, we are used to associating the ideas of sex and reproduction, there is no necessary connection between them, and it is reproduction, not sex, which is a precondition for evolution. Indeed, at the most fundamental level, sex and reproduction are exact opposites. In reproduction, one cell turns into two, whereas the essential feature of the sexual process is that two cells fuse to form one. Thus sex is an interruption of reproduction. Since, other things being equal, natural selection favours those types which reproduce most rapidly, there are real difficulties in giving a selective explanation for the widespread occurrence of sexual fusion. These difficulties are still unresolved, although many solutions have been suggested.

The consequence of sexual fusion is to bring together in a single cell genetic material derived from two different ancestors. As we shall see, there are several processes, in addition to the sexual process of higher organisms, which have this effect. Almost certainly, the most ancient process, and one which continues to play a central role today, is genetic recombination, illustrated in Figure 5.

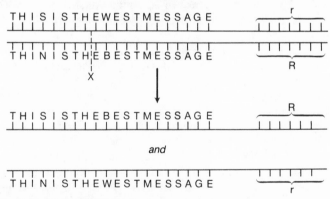

Figure 5. Recombination.

The base pairs in DNA molecules are represented by letters. In the upper diagram, both molecules contain 'errors'; that is, changes that reduce fitness. In the process of recombination, both molecules break at corresponding points, X, and the broken ends rejoin after changing partners, to produce the two molecules shown below. This process requires the presence of enzymes, to break and rejoin the DNA chains. R symbolizes a gene coding for such an enzyme, and r a length of DNA which does not code for an enzyme (or which codes for a less effective one). The result of recombination is to attach R to a perfect message; notice, however, that if initially the positions of R and r had been reversed, the result of recombination would have been to leave R attached to a message with two errors.

In that figure I have shown, in addition to the actual recombination event itself, a gene R whose presence is necessary if the event is to take place. The reason for including R in the picture is that, at least in all existing organisms, successful recombination requires the presence of enzymes, which in turn must be coded for by genes; it is such a gene that is symbolized by R. The symbol r in Figure 5 represents a gene which is ineffective in bringing about recombination. The problem of the evolution of recombination, and to a great extent of sex itself, is to explain why genes such as R have been favoured by natural selection.

There are two reasons why element R might be favoured, one immediate and one long-term. The immediate advantage is illustrated in Figure 5. We suppose that the nucleic acid molecule which survives and replicates best (i.e. the 'fittest') carries a message which is represented as THIS IS THE BEST MESSAGE. Any change in this will lower fitness. The figure shows that, if two messages each carry a different error, recombination can produce a copy without errors. The element R which made recombination possible has an evens chance of emerging attached to a perfect message. It has an equal chance of emerging attached to a message with two errors. However, if one error is fatal, two cannot be worse, so by causing a recombination element R has given itself an evens chance of survival, whereas without recombination it had none.

Historically, I have no doubt that this immediate advantage was responsible for the origin of recombination. Indeed, recombination probably emerged very early indeed in the history of life, and it spread because it was a way of repairing damaged nucleic acids. Genetic recombination is analogous to the process of producing one functional motor car by combining the engine from one broken-down car with the gearbox from another. However, as well as this immediate advantage, there is also a long-term one, which arises because at least some changes in the message are advantageous: there are beneficial mutations as well as harmful ones. The significance of this is shown in Figure 6. The essential point is that if two beneficial mutations, A and B, occur in different individuals in the same population, sex and recombination can bring them together in a single descendant. To stick to the motor car analogy, the design of cars has improved more rapidly because a designer can incorporate in a single design a

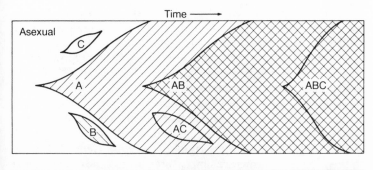

Figure 6. Evolution in sexual and asexual populations, after a diagram first used by the American geneticist H. J. Muller in 1932.

A, B, and C are different favourable mutations which, occurring in different individuals, can come together in a single descendant in a sexual population, but not in an asexual one.

synchromesh gearbox that originated in one model with fuel injection that originated in another. The consequence of this is that a population of individuals with sex can evolve faster to meet changing circumstances than one without it.

The significance of this long-term advantage is still being hotly debated. The crucial point to notice about it is that the entities whose 'fitnesses' are being compared are no longer individual organisms, but populations of organisms. We are saying that some populations (i.e. sexual ones) evolve faster than others, and hence will survive when others go extinct. Thus we are still using the Darwinian model of evolution by natural selection, but the entities to which we are applying it are populations, or species, and not individuals as in most applications of the idea. I shall return to this problem towards the end of this chapter.

First, however, I want to compare two ways in which we can look at evolution. One, the organism-centred way, is to see the individual organism as the centre of interest, since it is individuals upon which selection acts: individuals survive and reproduce, or they die. From this viewpoint, gametes and the genes they carry are merely a device whereby organisms produce offspring like themselves. The alternative is a gene's-eye view, suggested by Samuel Butler's famous remark that a chicken is simply the egg's way of producing another egg, and developed in a modern context in Richard Dawkins's book, *The Selfish Gene*. According to this view, organisms are simply survival machines, constructed by genes to ensure their own replication. It would be as foolish to argue about which of these views is correct as it would be to argue about whether algebra or geometry is the correct way to solve problems in science. It all depends on the problem you are trying to solve. However, in trying to understand the evolution of sex and recombination, I think the gene's-eye view is more illuminating. This is because we are concerned with the ways genes come together in groups, and then separate again. To be more specific, the gene R in Figure 5, which makes recombination possible, may have no effect at all on the survival of an organism in which it finds itself, or on the number of offspring produced; what R can affect is only the particular kinds of genes with which it will be associated in future generations.

To see the relevance of this, it is first necessary to know a little about the various ways in which genes are arranged in organisms. Until recently, the major division between living things was seen as being between animals and plants—a division based on the way the organism acquires energy. Today we draw the major division between 'prokaryotes'—i.e. bacteria and blue-green algae—and 'eukaryotes'—i.e. all the rest, including single-celled and multicellular animals and plants. There are many differences between these two groups, but for our present purposes two are critical:

i) The cells of eukaryotes have a nucleus, surrounded by a membrane, and containing a number of rod-shaped chromosomes. In prokaryotes there is no nucleus, and the single chromosome is in the form of a ring of DNA.

ii) In eukaryotes, the cytoplasm (i.e. that part of the cell outside the nucleus) contains various membranous structures, absent in prokaryotes, of which the most important are the mitochondria (where energy obtained from oxidation reactions is converted into a usable form—see p. 63), and, in plants, the chloroplasts (where light energy is trapped). Both mitochondria and chloroplasts contain DNA, and synthesize some, but not all, of their own proteins.

Prokaryotes originated over 3×10^9 years ago; eukaryotes are only about one-third as old. Hence it took about two-thirds of the total period since the origin of life to reach the eukaryote stage. The way in which eukaryotes evolved from prokaryotes is discussed later in this chapter. But first I must describe various genetic entities—viruses, phages, plasmids, and transposons—which are even simpler than bacteria. These entities have in common that they consist primarily of a length of nucleic acid (DNA or RNA), and can multiply only within living cells, prokaryotic or eukaryotic. They differ in whether they exist incorporated into the chromosomes or free in the cytoplasm of the cells they inhabit, and in how they pass from one cell to another. They can be grouped under three heads:

i) Viruses. These exist in an inert but infective state outside cells, as a core of DNA (or sometimes RNA) surrounded by a protein coat. If the nucleic acid core enters a living cell, it takes over the metabolism of that cell, so that the viral DNA is copied, and viral coat proteins, coded for by the genes of the virus, are synthesized. Typically, the cell is killed, and many new virus particles emerge. 'Bacteriophages', or 'phages' for short, are viruses that infect bacterial cells.

(ii) Plasmids do not construct for themselves special protein coats to enable them to survive outside their host cells, but can be transmitted from one bacterial cell to another if those cells come into contact. They ensure their own replication, but, instead of killing the cell they inhabit, they often code for proteins which help the host cell, for example by conferring resistance to drugs, or the ability to perform particular chemical reactions.

iii) Transposons. These are DNA elements which exist only as parts of chromosomes, or of plasmids. They differ from ordinary genes in that they can move from one site to another in the chromosome. They can multiply faster than the chromosomes, because, when they do move, a copy is left behind at the old site as well as a new copy appearing at the new one. Transposons found in prokaryotes usually code for one or more proteins that ensure their own replication. Similar elements have recently been found in eukaryotic cells,[1] but there is still much uncertainty about what they do, other than move about.

Although viruses, plasmids, and transposons are relatively simple, they are probably not primitive, because they can replicate only inside living cells, using some of the machinery of those cells. It is widely thought that they are descended from genes, or groups of genes, of higher organisms, which have somehow escaped and set up life on their own. Viruses are just intracellular parasites. The genes of viruses code for a protein coat, which helps them to survive the death of one cell and to infect another. If two similar viruses infect the same cell, their genes will recombine, and this may confer either the short-term or long-term advantages outlined above.

Transposons can also be thought of as parasites, although as yet we know little of the harm, or good, they may do to the host cell. They are the ultimate in 'selfish genes'. They have discovered that a chromosome is a good place to live if you want to get replicated. The mechanism whereby they move depends on a process of recombination, so they are parasites whose success depends on the devices that have evolved in cells to replicate and recombine DNA.

Plasmids are often symbionts rather than parasites: that is, they benefit rather than injure their hosts, for example by carrying with them properties such as drug resistance which may be of value to the host cell. Perhaps the most remarkable property conferred on a bacterium by a plasmid is the ability to behave in a sexual manner. A bacterium carrying certain types of plasmid can 'conjugate' with other similar bacteria. During conjugation, DNA is passed from one bacterium to another; this DNA may be the plasmid which caused the conjugation, or the bacterial chromosome itself, or both. We can therefore see such a plasmid as a temporary alliance between genes which induce conjugation in the host bacterium (and so make transmission of the plasmid itself possible), and genes which help the host to survive and multiply. I say that the alliance is temporary, because exchange of genes between plasmids is common.

I now turn to bacteria, the simplest organisms capable of a non-parasitic life. Some bacterial populations are asexual: no means of exchanging genetic material exists. This does not rule out evolution by mutation and selection. In other populations, appropriate conjugation-causing plasmids are present, and recombination between bacterial chromosomes does occur. Thus in bacteria, sex is under the control of an infective agent, a plasmid. From the viewpoint of the bacterium, plasmids have two effects. First, an infecting plasmid may bring with it some metabolic ability or resistance to some drug. Second, some plasmids confer on the bacterium the ability to transfer genetic material. The first of these effects is certainly beneficial, and the second may sometimes be; hence bacteria have not evolved special mechanisms to resist infection by plasmids, as they have to resist viruses.

There remains the question of how we should view a bacterial species—for example, the *Escherichia coli* popular with geneticists. The difficulty is best explained by contrast with a eukaryotic species—for example the house mouse, *Mus musculus*. I shall return to the nature of species in the next chapter. For the moment, the relevant point is that, in eukaryotes, evolution takes the form of a branching tree rather than a network of lineages that split and rejoin: once two lineages have split, they do not rejoin. If we knew enough and could identify all the individual animals alive, say, one hundred million years ago which were ancestral to existing mice, we would expect to find those animals all belonging to a single species (although, if we went back far enough, we might not call that species a house mouse). Similarly, if we could trace back the ancestry of all the genes in existing mice, through successive replications, for the same long period, we would expect to find those genes in animals belonging to a single species.[2] Thus there is a group of genes— mouse genes—which recombine with each other, but not with genes from unrelated animals. In saying this, we are denying the existence of really distant hybridization. When the ancestors of whales reinvaded the sea, they did not acquire fins by hybridizing with fish.

Now it is not clear whether the same picture is true of bacteria. We have seen that they can acquire new characteristics if infected by a plasmid. However, the genes brought in by a plasmid do not usually become a permanent part of the bacterium's own chromosome. But sometimes they can do. This raises the possibility that the genes present today in the chromosome of *E. coli* have not all been derived from the same ancestral population in the distant past—as have the genes of mice—but have come from very distantly related ancestors. If so, a bacterial chromosome would resemble a professional football team, whose players have been transferred into the club from all over the place.[3] We are not yet sure how far this is true. The evidence that has led people to think that it may be can be illustrated by an example. Some prokaryotes can 'fix' nitrogen; i.e. they can incorporate atmospheric nitrogen into amino acids. This is a complex process, requiring 15–20 separate genes. The genes in question seem to be very similar in prokaryotes which in other respects are not very alike, and which are thought not to be closely related. This suggests that the whole tool-kit of genes evolved just once, and has since been transferred from one kind of prokaryote to another.

It seems likely, then, that genes from distantly related prokaryotes are occasionally recombined. In eukaryotes, in contrast, recombination is usually confined to members of a sexually reproducing species, although there are some facts that suggest that distant gene transfer is not wholly absent.

Although, today, recombination in eukaryotes is in the main confined to fairly close relatives, the origin of the eukaryotes may have brought together genetic material from distant sources on a dramatic scale. It seems almost certain that the chloroplasts of higher plants (the intracellular organs of photosynthesis) are descended from free-living blue-green algae which became symbionts inside other cells. This is not implausible. There are a number of existing cases in which algae live inside animals: the animal carries the plant to sources of nutrients and light, and the plant synthesizes sugars which the animal can use. The reason for thinking that chloroplasts have this origin is that they still retain their own DNA and their own protein-synthesizing machinery. On similar grounds, it is thought that the mitochondria (see p. 63) are also descended from free-living prokaryotes which became symbiotic.

Once eukaryotes had evolved, it seems that opportunities for genetic exchange would have been severely limited. The difficulty was overcome by the evolution of the characteristic sexual cycle. Its essential feature is that there is an alternation of generations between a 'haploid' state (having a single set of chromosomes) and a 'diploid' state (having two sets). The production of haploid cells by the diploid requires a special series of cell divisions ('meiosis'), during which genetic exchange between the two sets of chromosomes takes place. The diploid form is then reconstituted by the fusion of two haploid gametes.

In some organisms the diploid form is the more complex and long-lasting. For example, in flowering plants the haploid stage is reduced to the pollen tube, derived from a haploid pollen grain that falls on the stigma of a flower and grows until it reaches and fertilizes the egg cell that lies in the flower's ovary. In higher animals the haploid stage is a single non-dividing cell, the egg or sperm. In other organisms, the haploid stage may predominate: for example, in mosses the conspicuous leafy structures are haploid, and the diploid form grows parasitically on the haploid plant. In still others—for example, some seaweeds—haploid and diploid stages are similar in structure and persistence. It is uncertain which of these two types of life-history is primitive. It is certain, however, that both higher animals and higher plants are diploid, with only brief haploid interludes. Why should this be so? There is no agreed answer. However, I think we are more likely to find the right answer if we ask the right question. We should not ask 'Why are structurally complex organisms diploid?' but 'Why do diploid organisms sometimes evolve greater structural complexity?' In other words, I am again recommending a gene's-eye view. In anthropomorphic terms, we should not ask why complex organisms have found it beneficial to be diploid, but why genes living in pairs should have found it beneficial to construct complex bodies to protect and replicate them. I suspect that the answer is along the following lines. Greater complexity of adult structure requires an increase in the amount of meaningful DNA. New DNA arises by making two copies of a piece of DNA which was previously present only once, and then altering the sequence, and hence the meaning, of one of the copies. It is easier for this to happen if there are in any case two copies of most genes present most of the time, as in diploids.

The origin of the sexual process remains one of the most difficult problems in biology. I cannot attempt to answer it here, but I can explain the difficulty. The major consequence of sex was to make genetic recombination possible, once the 'old-fashioned' prokaryote methods of plasmid transfer and conjugation had become ineffective. Genetic recombination, in turn, enormously expands the possibilities of evolutionary change, as was illustrated in Figure 6. But this is a long-term, prospective advantage, not an immediate one. Natural selection lacks foresight. A trait will not be selected merely because it will have, at some time in the future, beneficial effects. It is only present benefits that count.

To a biologist, males are organisms that produce small motile gametes (sperm), whereas females produce large gametes (eggs). Almost certainly, the first sexually reproducing organisms produced only small motile gametes, as many simple animals and plants do today. In effect, they were all males, although there was probably a division into 'plus' and 'minus' strains, such that gametes could only fuse if produced by strains of opposite type. Females, producing large, non-motile gametes, have evolved many times. The reason is fairly well understood. Once the adult is much larger than a sperm cell, it pays some individuals to produce a small number of large gametes, rather than a large number of small ones, because a large gamete (after fusing with a small one) has a much better chance of surviving to become an adult.[4]

Given that there are females, the problem of why sexual reproduction exists can be raised in a sharper form (see Figure 7). Suppose that, in a typical sexual species, a strain of parthenogenetic females arose, producing only daughters like themselves. We would expect the numbers of these females, relative to the typical sexual females, to double in every generation. Soon the whole species would consist only of parthenogenetic females. There are, in fact, many animal and plant species that have taken this road. Among insects, parthenogenetic species are quite common. Among vertebrates, no mammals, and only a few domestic strains of birds, are parthenogenetic, but there are wild 'species' of lizards consisting only of parthenogenetic females. But there is evidence that species which wholly abandon sex are short-lived on an evolutionary time-scale. It seems that parthenogenesis may establish itself because of its twofold immediate advantage, but that parthenogenetic populations are ultimately

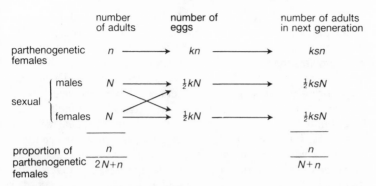

Figure 7. The twofold cost of sex.

It is assumed that each female, sexual or parthenogenetic, can lay k eggs, and that a proportion s of all eggs survive to become adults. The table shows how the relative proportions of the different kinds of females will change in one generation, starting with n parthenogenetic females, and N each of sexual males and females. If parthenogenetic females are rare (that is, n is small compared to N), they will double as a proportion of the population in each generation.

eliminated, probably because they cannot evolve rapidly enough to meet changing circumstances. However, there are also grounds for thinking that there must be short-term advantages to sex, large enough to counterbalance the twofold advantage of not having sons. Perhaps the strongest is that there are species in which some females reproduce sexually, and others by parthenogenesis: if there was not a short-term advantage to sex, the parthenogenetic females would long ago have replaced the sexual ones.

The picture which has emerged in this chapter is, perhaps inevitably, a confused one. It would be still more so if all the ways in which nucleic acid molecules can recombine had been included. It may help if I summarize the main conclusions. On the one hand, nucleic acids arise only as copies of pre-existing nucleic acids. In this sense, genetic information is potentially immortal. On the other hand, pieces of DNA are continually being separated and brought together in new combinations. In higher organisms the common process of recombination is the sexual one, and to that extent is confined to the DNA of a single species. Within the chromosomes of a single individual, however, there are processes whereby pieces of DNA are duplicated, or moved from place to place. Occasionally, even in eukaryotes, genetic material may move between distantly related organisms; in prokaryotes the possibilities of distant genetic transfer are much greater. The hardest questions are those concerned with the evolutionary causes, and consequences, of this pervasive genetic recombination. The questions are hard because a recombinational event may have no effect on the fitness of the individual in which it occurs, but only on what other genes a particular gene will associate with in future generations. The hardest question of all concerns the evolution of sex in eukaryotes, because of the twofold advantage to females that abandon it.

4

The pattern of nature

In the past two chapters, I have been discussing the mechanisms of evolution, but have said little about the outcome of the process. This chapter is concerned with the kinds of organisms we observe—the 'pattern of nature'—and the relation of this pattern to the evolutionary process.

It is now generally agreed that the natural way of classifying organisms is hierarchical: that is, in a series of nested sets. That is to say, we first classify organisms into species, and then group the species into genera, the genera into families, the families into orders, and so on. What is meant by saying that such a classification is 'natural'? After all, any collection of objects could be classified in this way. Thus suppose that various characteristics (size, shape, colour, etc.) were allotted to objects at random. The collection could then be classified hierarchically by choosing one character—say size—at random, and dividing the collection into orders according to that character. A second character—say colour—could be chosen randomly to subdivide each order into families, and so on, until a complete hierarchical classification was achieved. But there would be nothing particularly natural about such a classification. Apart from any other objection, a different classification would be reached if the characters were used in a different sequence.

This objection helps to reveal why it is that we regard a hierarchical classification as natural. In practice, unlike the example of a collection of objects with randomly assigned characteristics, we reach the same classification by using different sets of characters. For example, the vertebrates were first classified on the basis of their structure. If they are now classified afresh on the basis of biochemical characteristics, a very similar result is obtained. However, this only shows that there is some natural classification (which would not be the case if traits were

allotted to objects randomly), but not that the natural classification is hierarchical. There are other ways in which a set of objects
can be arranged. For example, the natural arrangement of the
chemical elements in Mendeleyev's periodic table has groups of
traits reappearing cyclically. It has not always been obvious that
the natural way to classify animals was hierarchical. Early
anatomists sought to arrange all living things linearly, on a single
'ladder of nature': man was about half-way up, with angels,
archangels, and powers above him, and a sequence of animals
below. In the early nineteenth century there was a group of
anatomists who held that the appropriate scheme was a set of
interlocking pentagons.

Hence it is not always true that the natural way of classifying a
set of objects is hierarchical; nor was it immediately obvious that
this is the best way of classifying living things. The justification
for using a hierarchy is mathematical, and has to do with the way
in which the total variability of a collection is successively reduced
as one passes from higher to lower levels in the hierarchy. For
our present purpose, however, the important point is that a
hierarchical pattern is what we expect to see if the objects being
classified have originated by a process of branching. The objects
need not be living things. Languages (but not motor cars) have
arisen by a branching process, and can be classified hierarchically.

Historically, people were not led to seek a hierarchical classification because of a belief in evolution; rather, they were pre-
judiced in favour of evolution because they had found a hierarchical classification to be appropriate. However, even if such a
classification is natural, it has arbitrary features. For example,
we place the small cats (domestic cat, sand cat, ocelot, etc.) in
one genus, *Felis*, and the large cats (lion, tiger, etc.) in a different
one, *Panthera*. It would be perfectly reasonable to keep them all
in the same genus. What would not be reasonable would be to
put the wolf into the genus *Felis*, while keeping the domestic dog
in the genus *Canis*. In other words, it is not arbitrary that the wolf
and dog are closer to one another than either is to the cat, but it is
to some extent arbitrary whether one recognizes some given
degree of similarity by placing two species in the same genus, or
in the same family. The pattern is not arbitrary, but the words we
use to describe it are, at least to some extent.

There is, however, one level of classification, the species, which is thought not to be wholly arbitrary. That is, it may be arbitrary whether one classifies lions as *Panthera leo* or *Felis leo*, but not whether one classifies all lions into one species or several. This is a much more debatable—and debated—point than appears at first sight.

As a first step in describing this debate, consider the positions taken up by the two leading naturalists at the end of the eighteenth century, the Swede Linnaeus and the Frenchman Buffon. Although both men were forced to modify their opinions later in life, initially they held the extreme 'nominalist' and 'essentialist' positions. Buffon held that only individuals are real. We group them into species merely as a convenience; if we did not do so, we could not give them names: hence 'nominalism'. For Linnaeus, in contrast, each species had its own essential characteristics—its essence; individual members of a species may differ, but only in non-essential ways.

At first sight, common observation seems to favour Linnaeus; animals and plants do fall into clearly distinct categories. If you buy a field guide to the birds or butterflies or mammals of an area, you will almost always find that the animals you see do clearly belong to one of the species described. In Britain, the tits you see will be blue tits, great tits, coal tits, crested tits, marsh tits, or willow tits (to be fair, you are unlikely to be able to distinguish the latter two unless you listen to their calls and songs). You will not see intermediates. The same is often true of plants, although it will be commoner to see plants intermediate between the described species. Perhaps the most convincing argument for the reality of species is that the kinds of animals and plants recognized by pre-literate peoples have been found to correspond almost exactly to the species recognized by modern taxonomists in the same area. If species were merely arbitrary groupings adopted to make naming possible, this would not be so.

If correct, the theory of the existence of discrete species, each with its own essence, brings biology closer to the physical sciences. Chemistry depends on the existence of the elements—oxygen, carbon, iron, etc. Any atom of oxygen is like any other in its essence, even if they differ in their position, velocity, state of excitation, and so on. (We now know that there are several isotopes of oxygen, but that does not seriously alter the argument.) Atoms of oxygen are quite unlike the atoms of any other element.

Although physicists are having some difficulty in deciding what
are the fundamental particles of which matter and energy are composed, they seem to agree that there are fundamental particles.
Thus both physics and chemistry are essentialist in outlook. Most
physical scientists are unable to imagine what their subject would
be like if there were no elements and no fundamental particles.

Yet despite common observation, and despite the example of
physics, the essentialist view of species has almost universally
been abandoned. This has not led us to adopt the nominalist
alternative. Instead, we hold that species are real things, but that
they do not have essences. This position, which would have
seemed contradictory to both Buffon and Linnaeus, requires
some explanation.

A species is a population of interbreeding individuals; more
precisely, because we do not want to regard the blue tits on the
Isle of Wight as belonging to a different species to those on the
mainland, a species is a group of actually or potentially interbreeding populations. The blue tits in any area may vary, but
they cannot be divided into two or more discrete groups (except
by sex, or by age), and the distinction between blue tits and great
tits in any area is absolute; intermediates do not exist because the
two species do not interbreed.

According to this view, interbreeding is at the same time the
criterion of whether two forms belong to the same species (e.g.
the dark and pale forms of the arctic skua interbreed freely, so
they belong to the same species), and also the reason why
organisms in nature do fall into discrete categories, with few
intermediates. The relevant criterion is whether two forms
commonly interbreed in the wild, and not whether they can be
persuaded to do so in captivity. For example, most ducks
belonging to the genus *Anas* (e.g. mallard, pintail, gadwall, teal)
will hybridize in captivity, and produce fertile hybrids. But they
do not do so in the wild, so the species remain distinct.[1] This at
once raises a difficulty. How are we to decide whether populations
living in different places belong to the same species? Sometimes
we can't. Thus if there is a geographical barrier, so that the two
populations never meet, it becomes an arbitrary matter whether
or not we group the populations into a single species.[2]

How do we know that sexual reproduction is responsible for

the relative uniformity of species, and that its absence is responsible for the discontinuities between them? There is, after all, the alternative theory that species represent different stable states of living matter, just as the chemical elements are different stable states of non-living matter. The critical test is to look at parthenogenetic populations, in which sexual reproduction has been lost. Sometimes, as in the various parthenogenetic 'species' of whiptail lizards, the populations are extremely uniform. However, this almost certainly indicates that each species has arisen in the relatively recent past (usually by hybridization between existing sexual species), and has not had time to evolve any substantial variability. In plants, however, there are some highly variable asexual populations. An example is *Hieracium* (Hawkweeds—yellow composites with flowers not unlike a dandelion's); the number of named 'species' in Britain runs into hundreds, but no two authorities agree on how many there should be. Thus in *Hieracium*, and in many other asexual groups, the plants growing in a region simply do not fall into a number of natural kinds.

Thus the pattern of variation in nature does not support the view that species represent a set of stable states, comparable to the chemical elements. Nevertheless, species are real things, with real discontinuities between them—at least if we confine ourselves to sexual organisms living in the same place at the same time. It is, in fact, the isolating mechanisms preventing hybridization between species that are the important feature. The problem of the origin of species is thus the problem of how such isolating mechanisms arose. There is in fact some disagreement about this question. One process which has certainly been important is geographical isolation. If two populations are separated by a geographical barrier, they are free to evolve in different directions. If isolation lasts for long enough, they may become so different that, if they meet again, they behave as distinct species. Some biologists regard this as the only, or at least as the predominant, mechanism of speciation, whereas others take a more pluralist view. All would agree that, particularly in plants, new species sometimes arise by hybridization between existing species. There is debate about whether a single species can split into two without geographical isolation. The difficulty with such 'sympatric' speciation is that species differ, not at a single gene locus, but by many genes; it is hard to see how such a difference could build up if sexual crossing was continuously breaking it down again. I do not think this difficulty is insuperable, particularly if one accepts that a single gene mutation could have a large effect in fitting an organism to a new environment. In fact, genetical analysis of plant 'ecotypes' (i.e. populations adapted to a particular habitat, for example growing close to the sea) shows that sometimes rather few gene mutations can cause most of the morphological change.

Species, then, are real, even if they do not represent different stable states of matter. What of higher categories—genera, orders, classes etc.? Pre-literate peoples not only agree with modern taxonomists about species, but in at least some cases agree in identifying the same higher categories. To give one example of a 'natural' higher category, consider the Anura, the order of amphibia which includes the frogs and toads. I do not think that anyone attempting to classify the vertebrates would fail to recognize the Anura as a group, or that, presented with an adult vertebrate, one would be in any doubt as to whether or not it was an Anuran. Of course, whether the Anura constitute a family, an order, or a class is an arbitrary matter: the point is that they do constitute a natural group.

It is harder to recognize the vertebrates as a natural group, and harder still for the molluscs (snails, clams, squids), because the similarity is of basic anatomy, and there are vast differences in shape and habits. However, no one today would question the validity of these groups. I should perhaps mention that there is some argument about the validity of such groups as the fish and the reptiles. Some taxonomists ('cladists') argue that we should only recognize 'monophyletic' groups: i.e. groups which include all the descendants of some common ancestor. The fish are not monophyletic in this sense, because the common ancestor of the shark and the herring was also the ancestor of all land vertebrates. The reptiles are not monophyletic either, because the common ancestor of lizards and turtles was also the ancestor of birds and mammals. Other taxonomists, while agreeing that the fish are not a monophyletic group, would nevertheless recognize them as a class, because they share many characteristics (e.g. the presence of gills in the adult) not present in other vertebrates. Although this issue has raised the passions of taxonomists, I cannot see that it is important. It concerns only how we name things, and not what we think is the case.

If we consider the largest groups, the phyla, such as the vertebrates, molluscs, or arthropods, there are no known intermediates between them, although they are all extremely similar in their genetic mechanisms, their biochemistry, and their cellular structure. In all probability, intermediates never did exist. That is, the common ancestor of these phyla was much simpler in its anatomy than any of them, and the anatomical complexities (e.g. segmented body plan, skeletal structures, image-forming eyes) evolved independently in the different phyla, although it is impossible to be sure of this in the absence of an adequate fossil record before about 600 million years ago.

The existence of these different anatomical plans, without intermediates, again raises the question of whether there are only a small number of possible body plans, represented by the existing phyla, or whether we are looking at the end results of a series of historical accidents. Pre-Darwinian comparative anatomists took the former view; since Darwin, the latter has prevailed.

My own view is of a series of historical accidents, subject to engineering constraints on the one hand, and to the conservatism of development on the other. Some examples will help to explain what I mean. The eyes of vertebrates and of squids are similar to each other, and to the camera. This is because there are rather few ways in which an image of the incident light can be formed. There is not only one way, as is demonstrated by the compound eyes of arthropods. Also, organic eyes are not subject to the same design constraints as are cameras: fish eyes have lenses made of material whose refractive index varies continuously from the centre to the outside. This is very efficient, but is at present beyond human manufacturing skills.

Thus anatomy may reflect the fact that there are only a few ways in which some engineering problem can be solved. However, resemblances of this kind are sometimes rather superficial. For example, flight has evolved four times (insects, pterosaurs, bats, birds). It has always involved thin movable extensions of the side of the body, but that is as far as the resemblance goes. Even for the three vertebrate cases, the ways in which the wing is constructed are quite different, although in each case they are modifications of the same fundamental structure, the pentadactyl limb.

These and other examples suggest that engineering design may impose some rather loose constraints on structure, and may in some cases impose precise quantitative constraints on shape (e.g. the shapes of bird's wings are very accurately adapted to particular modes of flight). However, the detailed structure of an organ can usually best be interpreted as a transformation of some ancestral structure. This leaves unanswered the problem of how and why the ancestral structure arose.

If we think about the basic body plan of a phylum, it can often be seen as a solution to an engineering design problem facing an ancestor, but one which is no longer relevant to most living descendants. Consider, for example, the vertebrates. They possess a segmented body musculature; a stiff axial rod, the notocord, or its replacement, the vertebral column; and a tail extending behind the anus. These features make it possible to swim by sinusoidal undulations. They possess two pairs of fins, or their derivatives, limbs: two is the minimum number of lateral extensions of the body needed to generate a vertical force through any point on the body (one pair could generate a force at only one point, and three pairs are one too many).[3] The structure of the skull, jaws, and gills of vertebrates has been interpreted, after a long and sophisticated study, as a filter-feeding device modified to take larger prey, fused to a bony box protecting the brain.

In other words, the body plan of existing vertebrates is readily understood as being derived from a set of structures which, in a distant ancestor, were adaptations for a particular way of life. Why, then, has the body plan remained the same when the way of life has changed? Men are not filter-feeders swimming sinusoidally in water. Yet human embryos still have a notocord. This suggests that the notocord still has some function in development, even if it no longer plays a role in swimming. Possibly the role is to induce the epidermis which it comes into contact with to fold up to form a nerve cord.

It seems to be a general feature of evolution that new functions are performed by organs which arise, not *de novo*, but as modifications of pre-existing organs. Our teeth are modified scales, our ear ossicles modified jaw-bones, our arms modified fins, mammae are modified sweat glands, and so on. If evolution proceeds by a series of small changes, it is hard to see how it could be otherwise.

The natural conclusion, then, is that the basic body plans of the different phyla represent structures which adapted some ancestral form to a particular way of life, and which have since been modified to serve different functions. Why, then, has the basic plan been preserved when habits and ways of life have altered? An engineer designing a horseless carriage is not obliged to retain structural features that existed solely to adapt the carriage to the horse (although, in fact, such primitive features were retained for a time). In evolution, structures are conservative because changes must be made one step at a time, each step being an improvement on the preceding one. Such stepwise change does not permit radical restructuring.

5

Problems of evolutionary biology

It is natural that people should question the theory of evolution by natural selection. No one wants to think that they are the product of blind chance and mindless selection. There is also a difficulty in conceiving that the observed complexities of animals and plants could have arisen in this way. Even Darwin admitted that his confidence was tested when he contemplated the vertebrate eye, although he went on to say that this was more a failure of the imagination than of the intellect. In this chapter, I discuss a range of problems that arise in applying Darwin's ideas.

i. Has there been time?

In Darwin's lifetime, physicists did not know about radioactivity, and therefore thought that the earth would cool down rather quickly. The length of time for which the earth had been at a temperature suitable for life was estimated at not more than a million years. For Darwin, this estimate was embarrassingly short, although fortunately he did not take it too seriously. The earth is now thought to be over 5,000 million years old, and the oldest fossils are over 3,000 million years old. Is this long enough for the evolution of an organism as complex as man?

This is a quantitative question, and demands a quantitative answer, however crude. One calculation, offered by opponents of Darwin's theory, is based on the improbability of an object as complex as a living organism arising by chance. Although fallacious, it is worth giving in more detail. Consider a small protein 100 amino acids long; if that is too improbable to arise by chance, then certainly a whole organism is. Since there are 20 kinds of amino acids, the number of possible proteins of that size is 20^{100}. This is an immense number. If the surface of the earth was covered to a depth of one metre in protein molecules, each one different from every other one, and if each molecule changed into another one once a second, and had done so since the origin of the earth, there would have been time to try out only a minute fraction of the possible sequences. Therefore, even if some particular sequence is the best possible protein for some particular function, and would be favoured by natural selection if once it arose, it could never arise in the first place merely by chance. It follows that some process other than random mutation and selection must be involved.

This argument comes up repeatedly: its latest manifestation is Hoyle's discussion of the likelihood of a wind blowing through a junkyard assembling a Boeing 707. What is wrong with it? Essentially, it is that no biologist imagines that complex structures arise in a single step. Even proteins formed as a random sequence of amino acids have some slight catalytic activity. If evolution by natural selection is to lead from some random sequence to one unique optimal sequence, there must be a series of intermediate sequences, between the random and the optimal, each being a slight improvement on the one before, the steps from one sequence to the next being a change in one amino acid, or at most of two or three amino acids. If this is so, then the random sequence can evolve into the optimal one in as few as 100 steps, each step being driven by natural selection. This could happen quite rapidly. An example of a unique sequence, of immense improbability, evolving from a random one in the laboratory will be discussed in the last chapter.

Implicit in this reply, of course, is the assumption that there is a sequence of functional intermediates between very simple structures, which could arise by chance, and the highly complex structures of today. I return to this point in section iii. of this chapter. First, I want to discuss whether there is any way in which we can estimate how long is required for evolution. I can see only one way in which this question can be approached quantitatively. This is to ask whether there has been time for natural selection to specify the DNA present in the human genome. Thus although we do not know how the DNA controls development, we are reasonably sure that it does carry almost all the information which has been produced by selection, and which is needed to control development. Although other structures are needed in the fertilized egg, only DNA has heredity, and hence can be programmed by selection.

There are approximately 10^9 base pairs in the human chromo-some set. A large fraction of this, however, is 'highly repetitive': that is, it is present in multiple copies, each with the same sequence. There is still much argument about what the role of this repetitive DNA may be, if any, but in any case natural selection did not have to programme each copy separately. It was sufficient to programme one copy, which was then replicated many times. If we take this into account, the total length of DNA to be specified is at most 10^8 base pairs.

How long would it take for selection to specify 10^8 base pairs? One rather crude way of looking at this is as follows. Suppose, in some population, a particular site in the DNA is completely random and unspecified: that is, it is equally likely to be C, G, A, or T. Suppose that individuals with A or T, say, are fitter than those with C or G, and that those with A are fitter than those with T. Full specification requires that all individuals have A at the relevant site. If natural selection eliminates half the population in each generation, complete specification requires two genera-tions of selection, one to eliminate C and G, and a second to eliminate T. Thus it takes two generations to specify each base. Since it is probably an exaggeration to suppose that as much as 50 per cent of the population dies selectively in each generation, let us be generous and allow ten generations per base. This suggests that $10 \times 10^8 = 10^9$ generations would be sufficient for selection to programme the human genome. The 3×10^9 years since the origin of life has been more than sufficient for this number of generations.

A more realistic approach would recognize that new DNA does not arise in evolution as a random set of sequences, but as a duplicate of some pre-existing and functional sequence. This would have two effects on the time required. First, the evolution of a new unique sequence with a new function may require the selective modification of as few as 5 per cent of the bases. This would reduce the time needed. However, those bases which were changed would not start with a frequency in the population of one in four, as assumed above, but with a low frequency generated by mutation. This would increase the time needed. The two effects would, very approximately, cancel out.

Thus there has been more than enough time for selection to generate specific DNA sequences of the required length. This conclusion is supported by the fact that changes in the fossil record, even rather rapid ones like the increase in human brain size in the last four million years, were slower by a factor of 1,000 than the rate at which changes can be produced by artificial selection in laboratory populations or domestic animals. Part of the explanation is that artificial selection changes only a few characteristics at a time, whereas in nature many changes occur simultaneously. Nevertheless, most natural populations, at most times, change much more slowly than they would if subjected to strong directional selection. In accord with this, studies of natural populations have shown that most selection weeds out extremes and preserves the status quo.

ii. Is all change adaptive?

There are many features of animals which could be improved on, and which are as they are because of the legacy of the past. For example, the bend in the lower spine of humans which is responsible for so many back pains exists because our ancestors were quadrupeds and we have only recently started standing upright. A more fundamental design fault is the positioning of our noses above our mouths, which requires our food and air passages to meet at the back of the throat, an arrangement which exists because the nostril in fish is not a breathing passage, but an opening to a chemical sense organ. This kind of maladaptive feature would be hard to explain if we had been designed by an all-wise creator, but is to be expected if structures change their functions in the course of evolution. In this section, however, I want to discuss the rather different possibility that some changes are essentially random.

The suggestion was made, some fifteen years ago, by the Japanese geneticist Motoo Kimura and by the Americans Jack King and Tom Jukes, primarily for changes at the molecular level. The idea is as follows. A protein consists of a string of from 100 to 500 amino acids. Occasionally, a change occurs in an individual which changes an amino acid. Such a change may be advantageous; if so, it will be established by selection as the standard in the population. Much more often, it will be harmful, and will be eliminated. But sometimes it will have no effect on fitness, or so little effect that its fate is decided by chance and not by selection. Such mutations are 'neutral'. According to the neutral theory of molecular evolution, most 'substitutions' of amino acids—that is, the replacement of one amino acid by another at some site as the standard type in the population—are neutral.

If we accept this neutral hypothesis, the rate of evolution for a given class of protein, for example haemoglobin, measured as amino acid substitutions per million years, will be roughly constant. The proof of this statement is so simple and elegant that it is the one piece of mathematics I have permitted myself in this book. Imagine, in some generation in the past, that a species contains N individuals, and hence $2N$ copies of some particular gene, say the gene for cytochrome C. Let the total mutation rate per gene be m (that is, the chance of a mutation of some kind occurring somewhere in the gene in a given sperm or egg). Most of these mutations will be harmful, and perhaps an occasional one will be favourable, but some fraction f are 'neutral'. Hence the number of new neutral mutations arising in the species per generation is $2Nmf$. Let us now leap into the future, and look at the genes present then. If we leap far enough, all the genes will be copies of one single gene in our ancestral population. Suppose that we look at a particular new mutant gene, out of the $2Nmf$ that arose. What is the chance that this particular mutant will be the ancestor of all the genes in the later population? Since it is neutral—that is, just as good or bad as all the other genes—the chance is $1/(2N)$. Hence the number of new neutral genes that arise in a generation, and later become fixed as the standard, is $2Nmf/2N=mf$. That is, the rate of substitution equals the neutral mutation rate.

If a protein is such that any change in its amino acid sequence will affect its functioning, then f will be small, and so will the rate of evolution. Hence, if the neutral theory is true, we expect those proteins with the most complex and precise functions to evolve slowest. There has been much controversy about the theory. It does seem to be true that proteins of a given kind do evolve at a surprisingly, although not exactly, uniform rate. However, the evidence that has finally convinced me that the neutral theory, if not the whole truth, is at least close to the truth concerns changes in the base sequence of DNA, and not in the amino acid sequence of proteins. There are various kinds of chromosomal DNA whose exact base sequence may well be largely irrelevant to their function—for example, because they are not translated into protein. It turns out that for these kinds of DNA the rate of change in evolution is much higher than it is for the more constrained, 'coding', regions. This is exactly what we would expect from the neutral theory.

One implication of the neutral theory is that we can use molecular changes as a kind of clock to measure evolutionary events. Thus suppose we have two existing species, and would like to know how long ago their ancestral lineage split into two. We can estimate this time by comparing molecular sequences in the two species. Such estimates are by no means certain. There is, however, one famous case in which it is beginning to look as if the molecular estimate is better than one based on fossils. This concerns the common ancestor of man and the great apes. The conventional palaeontological view was, until recently, of an age of 10 to 15 million years, whereas molecular data suggested a time of divergence of not more than 5 million years ago. Recent re-evaluation of the fossils suggests that the latter could be correct.

The neutral theory has been called 'non-Darwinian'. It is important to understand that it is in no sense 'anti-Darwinian'. Kimura has always accepted that morphological evolution is adaptive, and brought about by selection as Darwin proposed. He also accepts that some evolutionary changes in molecules do alter function, and are likewise selected. His point is that there are also molecular changes occurring which are not selected, because they have little or no effect on function. The view is still controversial, but I think it is winning.

iii. Does evolution always proceed uphill?

If Darwin was right, complex adaptations arise by a series of steps. Each of these steps must be an improvement, or, bearing in mind the last section, at least not a deterioration. Thus evolution can be thought of as a process of hill-climbing. Is this correct? Does evolution sometimes have to pass through intermediate forms of low fitness?

There is one fact that has led people to think that evolution has crossed valleys. This is that the hybrids between species are often—indeed, almost always—of low fitness compared to their parents. It does not follow, however, that the actual path of evolution involved a loss of fitness. By analogy, it may be possible to walk from one point in hilly country to another by a path which is always level or uphill, and yet a straight line between the points would cross a valley. Nevertheless, evolution may sometimes pass through intermediates of lower fitness. The only way in which this could happen is by chance in a small population: outsiders do come in at long odds.[1] But such transitions could only occur if the number of intermediate steps was small, and the fitness loss not too severe.

If evolution is hill-climbing, how can we explain the evolution of complex structures which would only work when perfected? The short answer is that we cannot, and that structures of this kind do not in fact evolve. The classic example is the eye. An image-forming eye, with lens, iris and so on, would not have evolved if nothing short of the formation of accurate images was of any use. But in fact we know of existing animals with a whole range of light-sensitive organs, from those which can merely detect whether the animal is in the light or the dark, through organs capable of detecting the direction of the incident light, up to proper image-forming eyes.

Often, the change in function of an organ during evolution is even more dramatic. Feathers, now used in flight, probably originated as heat insulators. The bone which in lizards conducts sound from the ear-drum to the inner ear originated as the hind border of a gill slit in a fish, and then became a strut bracing the jaw articulation to the skull, before it acquired its present function. Indeed, it seems to be a general rule that organisms acquire new capacities by modifying existing structures rather than by inventing wholly new ones. One of the evolutionary advantages of segmentation (the division of the body into a series of similar parts) is that it becomes possible to maintain an old function and acquire a new one simultaneously, by modifying the organs of one segment while leaving those of another unchanged. For example, the pincers of crabs and lobsters are modified walking legs.

iv. Are there 'group' adaptations?

There is no difficulty in explaining how a structure such as an eye or a feather contributes to survival and reproduction; the difficulty is in thinking of a series of steps by which it could have arisen.

I now turn to traits which seem not to contribute to the fitness of the individual even in their present form. One such trait—sexual reproduction—was discussed in Chapter 3. There, a major problem was whether the trait contributed to the fitness of the individual or of the population as a whole. In this section, I discuss other traits for which this difficulty arises.

It is a universal characteristic of higher animals that they senesce; that is, the likelihood of their dying in a given time interval, say a year, increases as they grow older, and often fecundity declines. Why should this be so? Surely, the longer an animal lives, and the longer it goes on producing offspring, the more genes it transmits to future generations? The question is interesting in part because of an erroneous answer that is sometimes given to it. Weismann, who was the first to raise so many important questions, suggested that animals senesce because, if they did not, there could be no successive replacement of individuals and hence no evolution. This is fallacious on two grounds. First, even if there were no intrinsic process of senescence, animals would die of accidents, and evolution would proceed. More important, if a trait is advantageous to an individual, it will become fixed in the species by selection, even if it is harmful to the long-term prospects of the species. Later in his life, Weismann saw the difficulty with this explanation, and offered other less vulnerable theories, one of which is not unlike that offered here.

In fact, it is not hard to think of explanations for senescence in terms of individual advantage. Perhaps an organism as complex as a man could never be wholly impervious to the ravages of time. I have little doubt, however, that we would live a great deal longer than we do, were it not the case that the same evolutionary changes that make an animal fit when it is young may condemn it to deteriorate when old. As a first example, consider our teeth. Our reptilian ancestors replaced their teeth continuously, as do lizards today. In contrast, we evolved a system with a single tooth replacement, of milk teeth by adult teeth. When our adult teeth wear out, we are in trouble. What was the advantage of not replacing teeth? Teeth which are permanent fixtures can evolve complex shapes, fitting precisely with their partners in the opposite jaw, making it possible to chew and slice in a way not open to most reptiles. The price you pay for being able to chew when you are young is that you will run out of teeth when you are old. This is rather a trivial example: if teeth were the only problem, senescence could be cured by false teeth. A more fundamental example concerns our brains, which are made of nerve cells which do not divide when we are adult. When a cell dies, it is not replaced. This places an upper limit on our lifespan. But a brain with a fixed number of cells almost certainly works better than one whose cells are constantly dividing. Now if there are changes, such as non-replacing teeth and non-dividing nerve cells, which make young animals more efficient, at the cost of condemning old animals to senescence, these changes will be favoured by natural selection, if only because most animals die of accidents anyway before they are old.

A second kind of apparently non-adaptive trait is self-sacrificing or 'altruistic' behaviour. The classic example is that of the sterile worker castes in the social insects, which sacrifice their own chances of reproduction in order to rear the offspring of the queen. How can such behaviour be favoured by selection? Since such altruistic acts are typically performed between members of the same species, and usually within a fairly stable social group, it is best to start by asking what causes animals to come together in groups. There are two kinds of reason, which we can call 'selfish herd' and 'mutual benefit'. The idea of a selfish herd is W. D. Hamilton's, and is perhaps best explained by his own imaginary example. He supposes that some frogs are sitting on the coping stones of a circular lily pond. In the pond there lives a grass snake, which will emerge at some random point on the surface, and eat whichever frog is closest to it. If you were a frog, where would you sit? Clearly, at whichever spot minimizes your chance of being eaten. The best you could do would be to sit very close to another frog, and if possible between two frogs which are close to one another. If all the frogs pursue this policy, they sit in a tight clump. Hamilton gave some entertaining examples of animals attempting to escape predation by clumping. The behaviour is entirely selfish. Nothing is done to prevent the snake from eating a frog: each frog merely does what it can to reduce its chances of being the unlucky one.

In contrast, if by grouping together animals can resist a predator altogether, as a herd of musk ox can drive off wolves, we can reasonably speak of mutual benefit. There are plenty of cases in which each member of a co-operating group is better off than it would be on its own. Such groups are likely to evolve more complex co-operative behaviour than will arise in a selfish herd. However, mutual benefit cannot explain self-sacrificing behaviour. In Darwinian terms, a sterile worker is not better off for being in a colony. Darwin himself was aware of the difficulty, and saw the kernel of the answer. He argued that families would do better if their members were altruistic. The matter was taken further by R. A. Fisher and J. B. S. Haldane, but our present grasp of the subject is largely due to Hamilton. His approach is essentially a gene-centred as opposed to an organism-centred one (see p. 30). A gene which causes its carrier to forgo reproduction in favour of raising relatives will increase in frequency in the population if the number of replicas of the gene in those relatives is greater than the number which would have been present in its own offspring. This process has been called 'kin selection'. If we want to know in precisely what circumstances such a gene would increase in frequency, we have to carry out calculations which have proved to be full of pitfalls for the unwary.

There are, however, two criticisms which have been levelled at Hamilton's ideas, mainly by sociologists and anthropologists, which are based on a misunderstanding. The first is that there cannot be a 'gene for altruistic behaviour', because such behaviour would require the action of genes at many loci, and also a host of environmental preconditions; behaviour is a product of the whole organism, not of a single gene. This is both true and irrelevant. If it were valid, it would apply equally well to the evolution of any complex trait whatever. If a geneticist speaks loosely of a 'gene for behaviour X', he does not mean that there is a gene which would cause behaviour X if inserted into any animal. He means that there is a gene which, if it is present in an animal of a given species, along with all the other genes and environmental circumstances characteristic of that species, will make that animal somewhat more likely to do X. Then, if doing X causes an increase in the number of copies of the gene in the population, the gene will spread.

The second misunderstanding concerns the ideas which must be present in an animal's mind if kin selection is to operate. By using the word 'altruistic' biologists do not wish to imply that animals are actuated by a concern for others, as will be obvious from the fact that they have written of altruistic behaviour not only in social insects but also in plants and even in viruses.[2] What is meant is that an organism does something which reduces its own chances of survival and reproduction, and increases the chances of other members of the species. A related misconception is that kin selection can operate only if an animal can recognize its degree of relationship to others. This is quite unnecessary. If, in a given species, animals at a particular stage of their lives habitually live with relatives, then selection will favour altruistic acts that cause a net increase in the number of the relevant genes, whether or not an animal can recognize its own relatives. As it happens, cases have been reported recently in which animals do seem to be able to recognize their relatives as such, but that is another story.

6

Stability and control

If you visit a biochemistry laboratory, you will often find on the wall a 'metabolic chart', usually supplied as an advertisement by a drug company, giving the names and chemical formulae of the thousand or more organic compounds commonly found in cells, connected by arrows indicating which substances can be directly transformed into which others. Most of the reactions shown are universal, occurring alike in bacterial cells, and in the cells of your liver and mine. Some familiarity with such a chart is part of the skill of being a biochemist, just as some knowledge of the anatomy and classification of animals is part of the skill of being a zoologist. I do not intend here to give any details of the substances present, or of the kinds of reaction which occur. There are, however, some general questions to which at least a partial answer can be given. Why do these particular reactions, and no others, take place in cells? Where does the energy come from to drive them? What controls the whole system? How does it come about that different substances are found in different parts of the body?

First, a few remarks about chemistry. The chemical elements are made of two kinds of particle, positively charged protons forming a central nucleus, and negatively charged electrons forming an outer shell. Since electrically charged objects attract or repel one another (unlike charges attract, like charges repel), there are forces of attraction and repulsion between atoms, which may cause them to combine in specific ways to form chemical compounds. For our present purpose, it is useful to distinguish two kinds of chemical reaction, according to whether the reaction releases energy or requires it. For example, when one oxygen and two hydrogen atoms combine to form a molecule of water, there is a release of energy. This energy appears as a vigorous vibration of the resulting molecule, which is experienced on the large scale as a rise in temperature. In order to split water into oxygen and hydrogen, we would have to supply energy.

It is not always true that combining two substances into one releases energy. For example, when two amino acids are joined, as in protein synthesis, energy must be supplied. How can such energy-requiring reactions take place? We can imagine the formation of such a chemical bond as requiring that a spring be compressed to bring the substances close to one another, at which point they are held together by a catch, as the jack is held in a jack-in-the-box. In order to compress the spring in the first place, the atoms must collide with sufficient energy. If the chemical bond is broken (that is, the catch is tripped), energy is released.

The energy levels shown in the two kinds of reaction are shown in Figure 8. Consider first Figure a1. Two substances, X and Y, combine to form a compound XY, and in so doing release a quantity of energy h. However, in order for the reaction to take

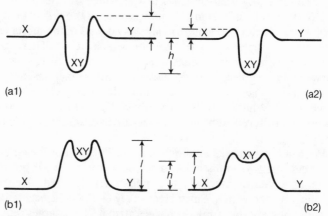

(a1)   (a2)

(b1)   (b2)

Figure 8. Energy levels in a chemical reaction.

(a1) represents a reaction in which a quantity of energy, h, is released when X and Y combine to form XY. The height l represents the 'activation energy' needed to start this reaction. To split XY into X and Y requires a net input of energy, h, and an activation energy $l + h$. (b1) represents a reaction in which a net input of energy, h, is required to combine X and Y to form XY, and in which the splitting of XY releases an equal quantity of energy. The activation energies in this case are l for the combining reaction, and $l - h$ for the splitting reaction. Figures (a2) and (b2) show the same two reactions when an enzyme has lowered the activation energy. Note that the enzyme does not alter the net energy change, h.

place, X and Y must collide with sufficient energy to overcome the 'activation energy', l. For this reason, a reaction may not proceed at low temperatures. Thus oxygen and hydrogen do not combine at room temperatures, but do so explosively if the temperature is raised. The effect of raising the temperature is to make the individual molecules move faster, so that collisions can overcome the activation energy.

For the energy levels of Figure a1, it is possible for the compound XY to break up into X and Y, but this requires the much higher activation energy, $h + l$. Therefore such collisions are rare compared to collisions in which X and Y combine, so that the overall effect is the formation of the compound XY from its components. Figure b1 shows the energy levels for a reaction, such as the joining of two amino acids, in which energy is required to combine X and Y, and is released when XY breaks up. Now the activation energy is l for the combining reaction, and $l - h$ for the splitting reaction, and the overall effect is for X and Y to split up.

With this rather crude model of chemical reactions, we can return to our first question: why do only certain particular reactions occur in cells? The answer is that a reaction will take place at an appreciable rate only if there is a specific enzyme that speeds it up. The reactions shown on the metabolic chart would proceed very slowly, or not at all, in the absence of enzymes. Enzymes are protein molecules whose function is to speed up chemical reactions: that is, they are catalysts. An enzyme works by combining reversibly with the substances whose reaction it is catalysing, and by so doing lowering the activation energy (Figure 8, a2 and b2). Notice that the enzyme can do nothing about the difference, h, between the initial and final energy levels. By analogy, if two cities are separated by a mountain, an engineer can, by digging a tunnel, reduce the height to which one must climb to travel between them, but if there is a difference of altitude between the cities he cannot alter it.

Every reaction on a metabolic chart is speeded up by an enzyme. It is easy to see how energy-releasing reactions can occur, but what about energy-requiring ones? Suppose that Figure 8, b1 and b2, represent the formation of a bond between two amino acids. The main effect of an enzyme, altering b1 to b2, would be to speed up the rate at which bonds were broken. To drive the reaction in the other direction, towards the formation of XY, requires a supply of energy, which must be very precisely applied. It is no use just heating things up; that would again only speed up the reaction in the wrong direction. To persevere with our mechanical analogy, if we picture the forming of a bond as equivalent to putting the jack in the box, we need to bring to the box a special tool with a ready-compressed spring, and couple them together in such a way that, as the spring in our tool is released, it compresses the spring of the jack and forces it into the box.

In this analogy, the molecule in the cell which corresponds to the tool is ATP, which is short for adenosine triphosphate, and the ready-compressed spring corresponds to the high energy phosphate bond in the ATP. When this bond is broken, a lot of energy is released, which can be used to drive other reactions. Clearly, it is no use breaking the phosphate bond unless the ATP molecule is in just the right place; otherwise the result would only be to heat things up. The ATP molecule is held in the right place by the enzyme which is catalysing the reaction. This explains why the tool is an ATP molecule, and not a simple inorganic phosphate; the energy is in the phosphate bond, but the adenine part of the molecule is like a handle by which the enzyme can take hold. That the handle is adenine, rather than some other organic molecule, is probably a historical accident, but it is not an accident that the phosphate is attached to some characteristic molecule that the enzyme can recognize.

To summarize, the organic compounds found in cells are built up and broken down by enzymes. These are protein molecules, which owe their specificity to their amino acid sequence, and hence ultimately to the genes. Those reactions which require energy to drive them also require the presence of ATP. The breaking of a phosphate bond in ATP provides the energy. It has been said that we can regard ATP as a common currency which must be spent to provide energy whenever it is needed—to drive chemical reactions, to cause muscles to contract, or to pump substances across membranes.

It is therefore natural to ask where we get the money from. Every time an ATP molecule is used to provide energy, a phosphate bond is broken, and it is converted to ADP, or adenosine diphosphate. Unless there is some way of converting

ADP back to ATP, the cell would soon run out of energy. Obviously, this is a reaction that requires energy: there are no perpetual motion machines. In animals, ATP is synthesized in special intracellular organelles, the mitochondria (see p. 31), where organic compounds such as sugars are oxidized and the resulting energy used to convert ADP into ATP. These organic compounds, ultimately derived from food, are the necessary source of energy for animals. But again, it is no use just burning sugars: that would merely make things hot. It must be done in such a way that the released energy helps to form a phosphate bond. I fear that no analogy based on hooks and springs and handles would help to explain how this is done. The process is one of the most complex in biochemistry, and is still only partly understood.

Plants acquire their energy in a different way, although many details of the process are surprisingly similar. The source of energy is sunlight. The transformation is carried out by organelles called chloroplasts. The energy is first trapped by special pigments (chlorophylls), and then used to synthesize ATP. Some of the individual steps in the process are sufficiently similar to those going on in mitochondria to suggest a common origin.

We can get a somewhat clearer idea of how enzymes work if we distinguish between two kinds of chemical bond, which differ in the nature and magnitude of the forces holding the atoms together. Most of the reactions one learns about in elementary courses in chemistry, and also most of the reactions shown in a chart of metabolism, require the forming and breaking of 'covalent' bonds. There is however a second category, of non-covalent bonds, which differ in that the energies involved, and in particular the activation energies, are much lower. Such bonds are important in holding together two molecules whose surfaces fit accurately into one another. Because of the low activation energies, such reactions are often readily reversible at room temperature. In a protein, the bonds that link the amino acids in a linear string are covalent, and are not readily broken once formed. When the string folds up to form a globular structure, however, the bonds are not covalent, and the resulting structure is flexible. When an enzyme binds to its 'substrates' (that is, the substances whose reactions it catalyses), the bonds formed are again non-covalent. The fact that an enzyme will catalyse a reaction for one particular substrate, but not for another very similar one, depends on its surface having precisely the shape that will fit that substrate and no other. Once the substrate is bound to the surface of the enzyme, covalent bonds may be formed or broken.

I now turn to the third of my questions: what controls the whole system? This raises the prior question of whether we can usefully speak of control, except perhaps of systems controlled by people. It is characteristic of the metabolism of a cell that the relative amounts of the different substances remain roughly constant, and that if these proportions are altered by some outside disturbance, they tend to return to their initial values when the disturbance is removed. Is there any difference between this kind of stability, and the stability of a vortex, which was used as an analogy to a living system in the first chapter? After all, if one interferes with the flow of water in a vortex, this pattern too will be restored when the interference ceases. Why should we think that metabolism is controlled, but not the vortex? Certainly there is no one controlling it, as a man might control a power station by moving switches.

Yet I think there is a difference between the control of living systems and the stability of physical ones, which has nothing to do with the presence or absence of a controlling intelligence. The point is best made by looking at a simple inanimate, but man-made, system, which in this respect resembles a living and not a physical one. Consider a centrally-heated house. There will be a thermostat which can be set so that, if the temperature falls below 20°C, say, the heating is switched on, and if the temperature rises a few degrees higher, it is switched off. There are two points about this to notice. First, there is a recognizable 'sense organ', the thermostat. This sense organ translates information about temperature into electrical information, which in turn controls the boiler: in the same way, the ear translates information about vibrations in the air into messages in the auditory nerve. The second point is that a small energy input into the thermostat can release a large quantity of energy from the boiler.

There is nothing corresponding to a thermostat in a stable physical system such as a vortex. What, if anything, corresponds to the thermostat in the metabolic system? As one would expect, the control of a system as complex as that portrayed in a metabolic chart requires a large number of switches, of several different kinds. I shall describe only two kinds, but they are representative ones. The first is called 'feed-back inhibition'. Suppose that some product, P, is made from an initial substrate, S, via a number of intermediates: $S \rightarrow A \rightarrow B \rightarrow C \rightarrow D \rightarrow E \rightarrow P$, say. Here, each letter represents a metabolite, and each arrow an enzyme-catalysed step. In feed-back inhibition, the final product P acts to inhibit the first enzyme in the series: that is, the enzyme that catalyses the step $S \rightarrow A$. Such an enzyme is said to be 'allosteric': it can exist in two forms. In the absence of P, it has a shape which enables it to bind to S and convert it into A. However, a P molecule can also bind to the enzyme, typically at a site different from S. When it does so, it alters the shape of the enzyme so that it no longer binds to S. The binding of P to the enzyme is non-covalent, and hence readily reversible. Clearly, as the concentration of P rises, so will the proportion of enzyme molecules to which P is bound; hence the rate of conversion of S to A, and thence to P, will fall. If the substrate S can be used for something else, it is clearly efficient to control the first step in the synthesis, $S \rightarrow A$, rather than, say, the last one, $E \rightarrow P$, because in the latter case E would accumulate.

In this typical example, the allosteric enzyme can be thought of as a switch. Similar switches can be used in other ways. For example, substance P can be used to control not its own synthesis but some other pathway, just as a thermostat could be used to switch on the television instead of the boiler.

Switches of this kind maintain the stability of complex metabolic systems without altering the numbers or kinds of enzyme molecules present. Protein molecules themselves, however, are not immortal: they are broken down, and must be replaced. Hence some mechanism is required to control the number of enzyme molecules. Such a mechanism is particularly useful if there is a need to alter the kinds of proteins present to meet changing circumstances. The classic example was discovered by the French biologists François Jacob and Jacques Monod in 1959 in the bacterium *Escherichia coli* (see Figure 9). This bacterium has a battery of three genes (G_1, G_2, G_3 in the figure), which code for proteins (P_1, P_2, P_3) needed by the cell if it is to use galactosides (energy-rich compounds which may or may not be present in the surroundings). If there are no galactosides, these genes are

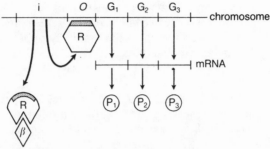

Figure 9. Gene control in a bacterium.

A gene, i, codes for a repressor protein, R, which binds to a special operator region of the chromosome, O, and thereby prevents the production of mRNA from the genes G_1, G_2, and G_3. If the substrate for the enzymes P_1, P_2, and P_3 is present (β in the diagram), it binds to the repressor, and changes its shape so that it can no longer bind to the operator segment, thus switching the genes on. The surface of the repressor that binds to the chromosome is shown shaded. Two repressor molecules are shown. One of them is bound to a substrate molecule, β, and in consequence its shaded surface is distorted so that it cannot bind to the chromosome. The effect of this control system is to ensure that the enzymes are produced when their substrate is present, but not when it is absent.

inactive and the proteins are not made: there is no point in making enzymes that cannot be used. Within a few minutes of galactosides being added to the medium, the genes are switched on, and enzymes appear.

As before, control depends on an allosteric protein, known as a repressor (R in the figure), coded for by a regulator gene(i). The regulator gene is always 'on', and synthesizes repressor at a very low rate. The repressor molecules bind very stably to an 'operator' region of the chromosome (O), and block the synthesis of messenger RNA from the genes G_1, G_2, and G_3. Hence no enzyme is made. Suppose, now, that galactosides are added to the medium. Molecules of galactoside (β in the figure) enter the cell, and bind to the repressor molecules. In so doing, they alter the shape of the repressor so that it can no longer bind to the operator region. As a result, genes G_1, G_2, and G_3 are switched on, and enzymes are produced. If galactosides disappear from the medium, the repressor molecules again stop the synthesis of enzymes. The two reactions of the repressor protein, with the operator region of the chromosome and with the inducing galactoside molecules, are non-covalent, and therefore readily reversible, so that synthesis can be switched on and off repeatedly.

Rapid reversibility is what is needed in a control system which must respond to environmental changes which may be transient, such as the presence or absence of galactosides. As I shall discuss in Chapter 9, more long-lasting changes in gene activation are involved in the differentiation of cells in higher organisms, for example the differences between cells in the kidney, liver, intestine and so on. The nature of these changes is less well understood than of those that occur in bacteria adapting to new chemical environments.

Monod, in his book *Chance and Necessity*, stressed one consequence of the methods of control just described. To quote, 'The result—and this is the essential point—is that so far as regulation through allosteric interaction is concerned, everything is possible.' Since there is no necessary chemical connection between the substances which bind to allosteric proteins, and the chemical reactions those proteins catalyse, it follows that the results of metabolism, although fully interpretable by the laws of chemistry, are not dictated by those laws, but by the physiological needs of the organism, and ultimately by natural selection. This property, which Monod calls 'gratuity', is analogous to the arbitrariness of the genetic code, discussed on p. 19.

The fourth question with which I started this chapter concerns the maintenance of a different chemical composition in different places. Even within a single cell, there are regions of different composition: for example, the mitochondria in which ATP is synthesized (see p. 63) are minute bags surrounded by membranes, with different substances present inside and outside. It is easier to start, however, with the simple fact that the chemical composition inside an organism is different from that outside. Consider a single-celled organism living in water. Clearly the various substances synthesized inside the cell must not be allowed to escape. However, the cell cannot be surrounded by an impermeable membrane, because it must acquire substances— 'food'—from outside. Nor is it possible to have a membrane with pores of just the right size to allow in needed substances from outside but not let essential substances escape, if only because the cell may need to retain some small molecules, and admit some larger ones.

Imagine, then, a cell surrounded by a membrane, and suppose that survival of the cell requires that some substance, X, be concentrated inside. If the membrane is permeable to X, then molecules will pass randomly across it, and the concentration of X will become the same inside and outside. There are two mechanisms which could accumulate X inside, both of which occur. In the first, there are protein molecules in the cell which bind specifically to X. Consequently, when an X molecule passes randomly into a cell, it may become bound to a protein, which is far too large to pass through the membrane. The concentration of X free in solution would still be the same inside and outside, but X would accumulate inside in the bound form. The second mechanism is to have special protein molecules in the membrane, able to take hold of molecules of X and pass them directionally from outside to inside. Such a process is called 'active transport'. I will discuss in a moment how such transport might be achieved, but first I give an example of its role in maintaining life.

The concentration of salts in the blood of fresh-water fish is much higher than it is in the water surrounding them. If they did nothing about it, water would flow in through their gills, and other permeable parts of their surface, and they would swell up and die. In fact, water does flow in, but it is eliminated again in their urine. In the kidney, body fluids are forced into the kidney tubules. Then salts are actively transported back out of the tubules into the blood, against the concentration gradient, leaving behind in the tubules a very dilute urine, which is voided to the outside. This is an over-simple account, but it does bring out the essential point that, at some stage, substances have to be transported across a membrane against a concentration gradient.

It is easy enough to assert that active transport across membranes occurs, but much harder to explain how. Clearly, the process requires energy. If, spontaneously, a system passes from state A to state B, it will require energy to drive it in the opposite direction. As expected, the energy is supplied by converting ATP to ADP in the membrane. Also as expected, the energy released is used by special proteins in the membrane. It seems that there are many such proteins, each concerned with the transport of specific substances. But I am afraid it would require a deeper understanding of the chemistry of proteins than I possess to explain how the energy of the ATP is harnessed to the active transport of salts and other substances.

I have discussed active transport at some length, because it is one of the processes on which the maintenance of spatial differentiation in organisms ultimately depends. Processes as diverse as salt regulation in fish, the acquisition of nutrients by plant roots, and the conduction of impulses by nerves, all depend on it. To give a last example, one of the proteins coded for by the genes illustrated in Figure 9 has the function of transporting galactosides into the cell.

At the scale of a single cell, molecules move from place to place by diffusion, although even at this scale larger objects like chromosomes and vacuoles are moved actively by the contraction of fibres called microtubules. In large multicellular organisms, diffusion would be too slow, and a system of tubes along which fluid can be pumped has evolved independently a number of times. At this scale, new problems arise, in addition to the biochemical ones just discussed. Essentially, these are engineering problems. Biologists try to interpret the structures they see as performing some function, subject to the constraints set by the materials of which they are made, and the conditions under which they must operate. Since I started as an engineer, and have ended as a biologist, I am struck by the similarities between the two ways of thinking, although there is the difference that an engineer starts with a function to be performed, and designs a structure to perform it, whereas a biologist often starts with a structure, and has to work out the function it is performing.

To illustrate this approach, let me give a very simple example. How is a land animal to solve the problem of acquiring oxygen without losing water? Any membrane which will permit the passage of oxygen will also permit the passage of water. Clearly, therefore, there is no way in which oxygen can be obtained without any water loss: the best that can be done is to minimize the loss. The only way of doing this is to bring a supply of air to an internal lung, and there to extract from it as much oxygen as possible. The extraction problem is one of biochemistry: its solution is the haemoglobin molecule, which will combine with any oxygen molecules which diffuse across the membrane lining the lung. Notice that this is an example of the first method I mentioned of concentrating substances on one side of a membrane: oxygen is withdrawn from solution by being bound to a haemoglobin molecule which is far too large to pass out across the lung membrane. The price paid is that the air in the lung is saturated with water vapour, which is lost when the animal breathes out (although some will be retained in the nasal passages if the air is cooled down before it leaves). The best that can be done, therefore, is to limit the supply of air to the minimum necessary to meet the requirements for oxygen. This can be done only if gas exchange takes place in an internal lung, to which air flow can be controlled. Contrast this arrangement with that in some frogs living in wet places, which have no problem with water loss, and breathe mostly through their skin, which is well supplied with blood-vessels. There remains the problem of controlling the flow of air. An animal in a desert must breathe as little as possible. Such control requires that the oxygen concentration in the blood be accurately measured, and that breathing rate be adjusted accordingly.

The parallel between this way of thinking, and that of an engineer designing an air-conditioning system, is obvious. The hard problems are often those of control and communication. In animals, there are two main methods of communication between different parts of the body. In hormonal communication, substances synthesized in one place are carried, by fluid flow, to another, where they act as chemical signals. In nervous communication, messages are carried as electrical impulses in nerve fibres. It is worth noting that the property of 'gratuity', discussed above in connection with allosteric enzymes, applies also to hormonal and nervous conduction. It is not chemically necessary that a particular hormone has a particular effect—for example, that adrenalin produces bodily changes associated with anger and fear: as far as chemistry is concerned, adrenalin could as well have evolved as a tranquillizer. In the same way, the significance of a particular sequence of impulses in a nerve fibre does not depend on the sequence itself: the meaning depends on where the nerve comes from (for example, what part of what sense organ), and its interpretation depends on where it goes to (that is, to what part of the brain).

To conclude this chapter, I would like to return to the picture, discussed in the first chapter, of the organism as a dissipative structure, maintained by the flow of energy through it. The picture is part of the truth, but only part. The maintenance of living structure requires not only a flow of energy, but a multitude of controls on that flow. At the smallest scale, allosteric enzymes act as sensory devices, measuring molecular concentrations and responding appropriately: that is, in a way that ensures the survival of the organism. The functioning of these enzymes depends on their amino acid sequence, and that in turn depends on base sequence in DNA, which is the result of many millions of years of natural selection. These controlling structures—proteins, and particularly DNA—are relatively long-lasting, and, in the case of DNA, their structure can be transmitted in the process of replication. Organisms, then, are like vortices in that their structure is maintained by a flow of energy, but the complexity of that structure is controlled by stable replicating molecules.

7

Behaviour

If this book were concerned only with the unsolved problems of biology, 90 per cent of it would be devoted to two topics: behaviour and development. In these fields, we are not even certain what kind of solution we are seeking. In the case of behaviour, we know quite a lot about what animals actually do, and about the anatomy and physiology of the brain, but only in the simplest cases can we explain the former in terms of the latter. In this chapter, I first describe three approaches to the study of behaviour: the analysis of animal orientation, behaviourism, and ethology. These three movements tackle different phenomena, although there is a striking similarity between the philosophies of the first two. I then discuss various ways in which we can think about the relation between brain and behaviour. One way is to ask what kinds of mechanism could in principle generate some observed behaviour. I illustrate this approach by discussing the problem of how animals find their way about. However, speculation about conceivable mechanisms is likely to be helped by physiological and anatomical studies. In the next chapter, therefore, I review some of the things we know about brains.

Curiosity about animal behaviour was greatly stimulated by the publication of *The Origin of Species* (1859). If, as implied although not explicitly stated in the *Origin*, man was descended from other animals, it was no longer possible, or so it seemed, to regard human behaviour as the product of the will, and at the same time to accept Descartes's view that animals are machines. Darwin himself was quick to appreciate this. Soon after the return of the *Beagle*, when he was still seeking for a mechanism for evolution, he began a series of notebooks, the M and N notebooks, which are in effect a search for à materialist theory of psychology: they are concerned with anything—for example, the effects of drugs, of ageing—which suggests that thoughts are influenced by the state of the brain. If natural selection was to account for the evolution of man's higher faculties, as Darwin was convinced it could, then those faculties must have a material basis. However, in the period following the *Origin*, the continuity between man and animals led others to seek explanations of animal behaviour in anthropomorphic terms; that is, in terms of wishes and intentions. At that time, these seemed to be the only alternatives: either men were machines, or animals were motivated by will and intention.

As a reaction against anthropomorphic interpretations, there grew up a movement devoted to explaining animal behaviour in mechanistic terms, a movement which was later extended to human behaviour. After all, the continuity between man and animal can as well be achieved by showing that man is a machine as by ascribing intentions to animals. I shall argue later that this contradiction between mechanistic determinism and free will is a misunderstanding. However, it seemed real enough in 1870. In consequence, the sciences of animal behaviour and experimental psychology were founded by men deeply hostile to anthropomorphic explanations.

The first serious attempt at a mechanistic account of behaviour was concerned with animal orientation. As a successful example of the approach, consider the following study of *Dendrocoelum*, a flatworm living in fresh water and usually found under stones. If you put some of these planarians in a dish illuminated vertically from above, and arrange that the intensity of illumination is greater on one side of the dish, the worms will congregate on the darker side. It is understandable, in terms of natural selection, that an animal whose natural habitat is the underside of stones should congregate in the dark: but how do they do it? Clearly they cannot be detecting the direction of the rays of light and moving away from the source, because illumination is from above. It turns out that the worms move at a rate which is unaffected by the intensity of illumination, but they change direction, to left or right, at a rate which does vary with light intensity. If the light intensity increases, the rate of turning also increases for a time, before falling back to its initial level. This simple reaction to light results in an animal spending most of its time on the dark side of the dish.[1] An explanation has been given for aggregation in the dark in terms of a simple behavioural response.

Curiously, this engineering approach had its origin in the study
of plant growth. Early in the nineteenth century, the tendencies,
or 'tropisms', of plants to grow towards light and against gravity
had been described. The founding father of the mechanistic
approach to behaviour, Jacques Loeb, adopted the word tropism
in his first book, published in 1910. The plant analogy guaranteed
freedom from anthropomorphism. For Loeb, it had the un-
fortunate result of committing him to too narrow a view of the
mechanisms responsible for animal orientation. There is more
than one way in which animals can congregate in the dark, or in
the light. For example, maggots move directly away from a light
source. As they crawl, they wave their heads from side to side. If,
in such a movement, the head meets a brighter light on one side
than the other, the animal turns away from the brighter side. As
it happens, neither this mechanism, nor that found in *Dendrocoelum,* bears much resemblance, even by analogy, to plant
tropisms, or to the mechanism Loeb believed to underlie animal
orientation. However, what was important about Loeb was his
commitment to mechanistic explanations, rather than his narrow
and largely mistaken choice of a particular mechanism. It is
characteristic that he stimulated J. H. Hammond to construct a
machine which oriented towards light. He wrote of this machine
in 1912: 'there is no more reason to ascribe the heliotropic
reactions of lower animals to any form of sensation, e.g. of
brightness or colour or pleasure or curiosity, than there is to
ascribe the heliotropic reactions of Mr Hammond's machine to
such sensations.'

Once it was recognized that animals orient themselves in a
variety of ways, the engineering approach was extremely successful in identifying the underlying mechanisms. As far as reactions to
simple stimuli such as light, gravity, temperature, and humidity
are concerned, the work was largely completed by 1940. However,
it provided the basis for further advances, including the discovery
that some animals can orient to previously unsuspected stimuli
(magnetic and electric fields), and von Frisch's discovery that bees
can communicate information about direction symbolically.[2] A
more fundamental change has come from studying how animals
orient themselves not in a simple environment but in a structurally
complex one; for example, how they use landmarks to find their
way home. I shall return to such problems later.

The movements analysed by Loeb and his successors did not depend on previous experience. Much of animal behaviour, however, is modified and moulded by experience. Serious work on learning originated with I. P. Pavlov in Russia, and with the behaviourists J. B. Watson and B. F. Skinner in the United States. To Pavlov we owe the concept of a 'conditioned reflex'. An unconditioned reflex is one which appears without previous experience; for example, if food is placed in a dog's mouth, it salivates. Pavlov found that if some other stimulus—for example, the ringing of a bell—immediately preceded the food, the dog would salivate in response to the ringing of the bell. Such a response is a conditioned reflex.

There is an obvious difficulty in explaining all learnt behaviour in terms of conditioned reflexes. Animals learn to do many things that they never do as an unconditioned response to any stimulus. For example, a dog can be taught to jump through a hoop, but there is no stimulus to which jumping through a hoop is the unconditioned response. This difficulty was overcome by the concept of 'operant conditioning', developed by the behaviourist, Skinner. The idea is that if some action, X, is followed by a stimulus, for example food, which is 'reinforcing', then the animal is more likely to do X in the future. The previous sentence could be taken as a definition of the word 'reinforcing'. However, it is a fact that there are stimuli such as food which are reinforcing in this sense. Clearly, if an animal is to learn to do X by operant conditioning, it must occasionally do X in the first place; otherwise X could never be reinforced. However, X does not appear in the first place as a response to some specific stimulus (as in the case of a reflex), but more or less spontaneously. In effect, an animal tries out a range of actions, and repeats those which are reinforced.

There is an obvious analogy between operant conditioning and evolution by natural selection. Behaviour becomes adapted to the environment by the reinforcement of spontaneous acts, just as morphological structure is adapted by the natural selection of spontaneous mutations. There is also a causal connection between the two processes. For operant conditioning to work, at least some stimuli must be reinforcing, and others aversive, without previous training. Otherwise, there is no way in which the process of learning can get started. Food is reinforcing, and injury aversive, because past natural selection has favoured animals in which this was so.

I have described the process of operant conditioning discovered by the behaviourists, but not the philosophy that guided them. In many ways, their philosophy was similar to that which guided Loeb and his successors. This was not accidental; Loeb had moved from Germany to Chicago, where he influenced the young J. B. Watson, who was just starting his research career. Like Loeb, Watson and Skinner were reacting against the mentalist concept that behaviour is caused by feelings, ideas, wishes, or intentions. Their basic objection was that they found such concepts unhelpful and non-explanatory. For rather similar reasons, Darwin reacted against Lamarck's idea that organisms have an inner drive to evolve greater complexity. To say that a dog eats because it wants to is as uninformative as to say that dinosaurs got larger because they had an inner drive to do so. The behaviourists insist that we can observe two things: first, what an animal does, and second, its previous environmental history. We can deduce that what an animal does depends on its state, and that its state depends on its past history of reinforcement and on its genetic propensities (for example, its propensity to be reinforced by food). It adds nothing to our understanding, Skinner argues, to talk of its frame of mind. Of course, what applies to animals applies equally to people: indeed, it has always been the declared aim of the behaviourists to understand and to control human behaviour. The basic view is best expressed by Skinner himself: 'A person disposed to act because he has been reinforced for acting may feel the conditions of his body at such a time and call it "felt purpose", but what behaviourism rejects is the causal efficacy of that feeling.'

The similarity in philosophy between Loeb and Skinner is apparent. The ethologists had a very different approach, deriving from a different set of questions and of methods. Both Konrad Lorenz and Nicko Tinbergen, and most of their followers, were led to a study of animal behaviour by a passion for natural history. There has consequently been a much greater emphasis in their work on the study of animals in the wild, on the differences between species, and on the adaptive significance of behaviour. This set of preoccupations led the ethologists to assume a more complex and extensive set of genetic endowments underlying behaviour. In particular, they argued for the existence of 'innate releasing mechanisms' and 'fixed action patterns'—concepts that are best explained by examples. The classic example of an innate releasing mechanism is afforded by herring gull chicks, which peck at their parent's bill to elicit food. In chicks without experience of an actual parent, the peck is best elicited by a thin downward-pointing projection with a red spot on it. To give a second example, pied flycatchers which have never seen a shrike will mob a model shrike; the mobbing is best elicited, or 'released', by a model of the right size and orientation, with a horizontal black eye stripe on the appropriate background colour. A brain structure responsible for such a response is an innate releasing mechanism.

Just as an animal without previous experience may respond to a complex stimulus, so it may produce a complex set of movements in response to a stimulus. An example is the sequence of acts performed by a mallard drake during courtship. The bill is shaken laterally, it is lowered into the water and raised suddenly to lift a plume of water, the bird rises vertically in the water, and it then settles again and shakes its tail from side to side. In the case of such a fixed action pattern, ethologists suppose that there is an innate structure in the brain, able to send out an appropriate sequence of commands to the muscles. Still more complex behaviours, such as the spinning of an orb web by a spider, or the singing of the species-specific song by a bird, depend on an interaction between innate structures generating particular movements, on innate tendencies to be reinforced by particular sensory inputs, and on experience, which may have to occur at particular times during development.

Ethologists have been particularly struck by the complexity of the behavioural responses of comparatively 'naïve' animals (that is, animals which have had little opportunity to learn). This has led them to postulate the existence of 'structures' in the animal's head: in contrast, behaviourists are reluctant to say anything about what is in an animal's head, other than that the members of a particular species have an innate tendency to be reinforced by some stimuli and not by others. The same disagreement has emerged more recently in the context of human language acquisition, in the debate between those who, like Noam Chomsky, suppose that human beings have an innate capacity to acquire language, and the behaviourists, who argue that language acquisition can be explained in the same way as most other behaviour, by operant conditioning: we learn to speak grammatically because we receive reinforcement when we do so.

These debates force us to ask ourselves what is the best way to think about behaviour and its relationship to the brain. I suggest that there are three possible approaches:

i) We can treat the brain as a black box into whose contents it is not efficient to enquire. We can observe the experiences that an animal has, and what it does, and study how the latter depends on the former. This in effect is the behaviourists' approach. They do not deny—indeed they insist—that what an animal does is influenced by past experiences as well as present ones. By implication, therefore, they accept that experience does alter the state of the brain. They also accept that the brain has certain innate dispositions, if only the disposition to be reinforced by particular stimuli. They deny, however, that anything useful is added by referring to states of the brain in terms of awareness or consciousness, or even as representations of the environment.

ii) We can take the lid off the box and poke around inside. This is the approach of the neurophysiologist. The difficulty is that, on a cursory inspection, what is inside the box is a grey mush, which, on examination under the microscope, turns out to consist of a confusingly large number of nerve cells (some 10^{11} in man), each connected to a large number of others. Despite this difficulty, any final account of how the brain works will have to be interpretable in terms of what its constituent nerve cells do. Some cases in which progress has been made are discussed in the next chapter.

iii) We can build 'models' of what we think might be in the box. If we know how something behaves, we can sometimes guess how it works. Let me give a classic example from another branch of biology. Mendel deduced, from the way in which characteristics appear in the progeny of crosses, that these characteristics were caused by 'factors' which obeyed certain rules (for example, that there are two factors in an individual, of which each gamete receives only one, at random). Forty years later these factors were identified as bits of chromosome, and a century later as molecules of DNA. In other words, Mendel's theory was a 'model' of what had to be there to generate the observed behaviour. It did not specify what the components of the model were made of, but only their formal relationships. By analogy, we can hope to deduce what must be in an animal's head by observing what it does. One reason for being hopeful is that we now make machines which do some of the things that brains do, although we must remember that there may be important ways in which brains and existing computers work on different principles.

As I see it, these three attitudes are not different theories about what the brain is like, but three different views about the best way to study it. Hence there can be no contradictions between their predictions which could be settled by experiment. The proper relationship between the neurophysiological and the model-building approach is clear enough: they should support one another. Given a clear model of how some behaviour is generated, it should be easier to seek the underlying physiological mechanism, just as, given Mendel's theory, it was easier to seek the material carriers of heredity. The support can work the other way; physiological knowledge can help us to build models.

The relationship between behaviourism and the other approaches is less clear. Originally, the behaviourists objected to explaining behaviour in terms of mentalist concepts, such as awareness or feeling, on the grounds that such 'explanations' explain nothing. However, it does not follow from this that there is nothing in an animal's head, or that nothing useful can be said about it. I think that the most fruitful approach is to ask what are the computational and cognitive capacities which we are forced to suppose that animals possess. I will try to illustrate this by discussing how animals find their way about. But first, something has to be said about consciousness and intentions, because in everyday life we interpret our own behaviour in these terms, and it seems natural to do the same for animals. The issue is philosophical, but it is so central to the analysis of behaviour that I do not think it can be avoided.

As I sit writing, I decide to go into the kitchen and make a cup of coffee. To me, it seems that my felt desire for coffee is the cause of my going to the kitchen. To a behaviourist, as the quotation from Skinner on p. 76 makes clear, this is an illusion. I go to the kitchen because coffee is reinforcing, and when I have gone to the kitchen previously I have been reinforced. I agree that I have been reinforced (although I'm not sure that I learnt how to make coffee by operant conditioning—I suspect I copied someone, or was told how to do it—but that is a separate issue). But that previous reinforcement is now represented in my brain: my brain is different because of it. It is that representation in my brain that is the immediate cause of my actions. Unfortunately, I can give no detailed physiological account of the ways in which my desire for coffee, and my knowledge of how to find the kitchen, boil a kettle, and so on, are actually represented. It is therefore natural to speak of my desire for coffee as the cause of my going to the kitchen; a quite different kind of cause would be operating if someone twisted my arm behind my back and forced me to go. But to say that my desire for coffee is the cause is not to deny that there are physical structures in my brain that are the cause, or that those structures are present because of my previous experience. The physical events in my brain, and the feelings that I have, are not alternative and mutually exclusive causes of my actions: they are simply different aspects of the same cause. Of course, I assume that you have feelings similar to mine, and I also assume that animals which are reasonably like us have feelings that are like ours.

The philosophical position I have just outlined is not like a scientific theory: it cannot be tested experimentally. Nor is it like a mathematical theorem: it cannot be deduced from a set of axioms. It is no more than a consistent point of view, which has the advantage that, if we accept it, we can stop arguing about whether feelings are causes of actions, and can get on with finding out how our brains work, without fearing that an answer to the question would make free will an illusion. Of course, a full acceptance of the point of view would rule out the continuation of thoughts and feelings when the brain stops functioning.[3]

I want now to turn to a less philosophical question about my passage from my armchair to the kitchen. How do I actually do it? Figure 10 shows an (imaginary) map of my house. A is the armchair and E the gas stove. The dotted line A–C–D–E represents the only sensible route. The task of going direct from A to C, if I can see C, presents no special difficulty; this is the kind of problem solved by the students of animal orientation. I solve it by pointing my eyes towards C, bringing my body into line with my eyes, and walking forwards. (The problem of recognizing C

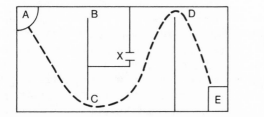

Figure 10. An imaginary map of my house.

Point A represents my armchair, and E the gas stove. The diagram on the right is a tree showing possible routes.

as the same object when viewed from different directions is a much harder one, which I will touch on in the next chapter). But why do I go to C, rather than to B? The behaviourists' answer would be that I have in the past been reinforced by coffee when I have gone to C, but not when I have gone to B. This could well be correct. It is certainly a simple explanation. All that has to be stored in my brain is 'If you want coffee, and are at A, go to C, and then D, and then E.'

There are, however, difficulties in explaining all behaviour in this way. In a larger building, like that in which I work, I would have to store a lot of instructions of the above kind in order to find my way about. A more serious difficulty is as follows. Suppose I go through door B (to the bathroom, say), and find that a new door has been constructed in position X. Next time I am in my armchair and want coffee, would I pursue the path A–B–X–D–E, which is a bit shorter than A–C–D–E? Probably not—I am a creature of habit. But suppose I was in the bathroom, and wanted to go to the kitchen. Would I go B–C–D–E, or X–D–E? I think that even I might manage the latter, even though going through door X had never previously been rewarded.

In order to solve this problem, an animal would have to store information in a different way. It would not be sufficient to store a set of instructions on how to act. It seems necessary for it to have what has been called a 'cognitive map' in its head; that is, it would have to have something which conveyed the same information as is conveyed by Figure 10. The word 'cognitive' is critical. The implication is that, to solve some kinds of problem, an animal must know that something is the case, and not merely that a given action or sequence of actions has in the past been reinforced. In this particular example the knowledge concerns the arrangement of things in space—hence the word 'map'.

There is, surprisingly, still a good deal of uncertainty about whether animals—and if so, which animals—can solve problems of this kind. Consider first a case in which no cognitive map is needed. Jumping spiders will make detours to reach their prey; for example, it may be necessary for the spider to leave the branch on which it is standing, climb up the stem, and walk out along another branch. It does so by going to a series of intermediate goals, for example the junction of the branch it is on with the main stem. While doing so, it remembers the direction of its prey in three dimensions, allowing for the effects of its own transverse movements. This is a fairly sophisticated performance, but it does not require a map. It is sufficient for the spider to follow a series of (probably innate) rules; for example, 'if you cannot proceed directly towards the prey, go to a junction and try again'.

Now consider an ingenious experiment[4] which almost forces us to suppose that an animal has a cognitive map. Rats placed in opaque water learnt to swim towards a submerged platform which they could not see. After as few as six trials, a rat placed in the water would swim almost directly to the platform. If, during training, the rat was always released in the same place, and if it was then released in a different one, it could still swim directly to the platform. This suggests that the rat has learnt that there is a submerged platform in a particular location relative to the walls of the pool (and to objects visible outside the pool), and not merely some rule such as 'when put in the water, swim in a given direction relative to the wall'. This kind of ability can be important in nature. An entertaining example of animals solving the same problem as the swimming rats, but with land and water reversed, is the ability of a tide-pool fish, the goby, to jump from one pool to another without landing on the rock between. It learns the relative positions of the pools at high tide, when it can swim over them.

Of course, the statement that an animal can solve a problem because it has a cognitive map is not the end of the matter. It only raises a further set of questions. How is the map formed in the first place? Where and how is it stored in the brain? These questions have not been answered, although some things are known. For example, it seems that rats must be able to move about and explore their environment if they are to form an adequate map of it.

Now consider what may seem a much easier question. Granted that I have a map of my house in my head, how do I use it to go from A to E? You might answer, 'Given the map in Figure 10, I can see at once that the dotted line is the best route.' This will not do, for two reasons. First, it is not clear what it would mean to say that I can see the map in my head. What would I be looking at it with? Second, people can solve problems without knowing how they do it. For example, I used to play squash reasonably well, although I have only monocular vision. The calculations I must have performed to decide where the ball was going to be, and to hit it, are staggering, but I managed (so long as I did not stop to ask how I was doing it).[5]

What kind of answer do I want to the question of how I use a map in my head? People are increasingly coming to think that we want an answer which can be simulated on a computer. That is, we want a computer program which, given the map and asked 'how should I go from A to E?' will answer 'A–C–D–E'. It would probably proceed as follows: from A, go to all the openings you can see; from each of those openings, go to any further openings you can see; by continuing this procedure, generate the tree of Figure 10; report the shortest sequence starting at A and ending at E. The reason for preferring a computer program to a verbal account (for example, the one just given), is that, when writing a program, you are forced to say exactly what you mean; if the program works, you know that there are no hidden difficulties with your explanation. For example, in writing the program outlined above, you would have some difficulty in telling the computer what an 'opening' is.

In discussing how animals find their way about, I have drawn a distinction between cases (for example, jumping spiders making detours) in which what is stored in the brain is a set of rules to be followed—sometimes called an 'algorithm'—and others (for example, rats finding a submerged platform) in which the brain stores a representation of the world, or, if you prefer, in which the animal has some knowledge of what is the case. Computer programs store knowledge in both these ways. Of course, the fact that you have written a computer program that will do what an animal does is no proof that the animal does it in the same way. By this, I do not mean that the animal uses neurones, whereas the computer uses transistors. I am concerned with the logical structure of the solution, and not with the material objects that embody that structure: in computer jargon, I am concerned with software, not hardware.

Given that there are several ways of doing a calculation, we can ask which of them, if any, is analogous to the way the brain performs the same calculation. An example will make this clearer. You wish to calculate the distance you would have to walk to get from London to Brighton. I provide you with a one inch to the mile map, and a box of matches each one inch long. One method is to lay the matches end to end on the map between the two towns, and then count the matches. Alternatively, you look at the map references of the two towns on a one mile grid; say they are (63,27) and (103,57). You then use Pythagoras' theorem to calculate the distance: that is, $((103 - 63)^2 + (57 - 27)^2)^{1/2} = 50$ miles. The first of these two methods is analogous to the real-world process of walking from London to Brighton and counting the steps; the second is not.

How could the brain perform such a calculation? One possibility is as follows. Suppose that a cognitive map, similar to that in Figure 10, exists in the brain. Further, assume that the map consists of a sheet of neurones, with particular neurones corresponding to particular places, the pattern being geometrically similar to the environment being mapped. (These are assumptions: a 'map' could consist of a list of places, each with a map reference.) Suppose further that each neurone forms synapses (conducting junctions) with its neighbours, but not with distant neurones. The distance between any two points would then be proportional to the number of neurones a message must traverse to get from one to the other; it would also be roughly proportional to the time taken for a neural message to travel between them.

If either of these methods were used to estimate the distance, the process would be analogous to laying matches end to end, or to walking and counting steps. The example is imaginary. We do not know how cognitive maps are stored in the brain. However, given that such maps exist, I find it hard to believe that distances could be estimated by a method analogous to using Pythagoras' theorem. A computer programmer would almost certainly get a computer to do it that way, but I doubt whether the brain does.

Occasionally, we can get some insight into how calculations are performed from behavioural data. For example, subjects (which can be people or, with suitable techniques, animals) are asked whether two complex shapes, presented in different orientations, are or are not the same. The time taken to give a correct affirmative answer increases linearly with the angle through which one shape must be rotated in order to coincide with the other. It is hard to explain this unless, in the subject's brain, the representation of one shape is rotated until it matches the representation of the other.

I will now try to summarize what I have said about how an animal finds its way about. Suppose it wants to go from A to a point E that it cannot see. There are two ways of achieving this. One is to follow a set of rules—an algorithm—such as 'go to B, then to D, and then to E'. The other is to build up a cognitive map of the region to be traversed. There is no doubt that the former method is often used. The relevant algorithms can be learnt by operant conditioning, as proposed by the behaviourists. There are, however, cases which are hard to explain in this way; examples are the ability of a rat to find a submerged support, and of a goby to leap to a pool it cannot see. These cases suggest that animals can form cognitive maps. However, this leaves much unexplained: how is the map formed, and how is it used?

In answering such questions, computer simulation is helpful, because you cannot simulate something you do not understand. The snag is that there may be more than one way of carrying out a task; the example I gave was measuring a distance by arranging matches end to end, and by using Pythagoras. One method may be analogous to the way the brain does it, and the other not. Sometimes we can get a clue from behavioural data; an example is the evidence that we compare shapes by rotating their images in our heads. Another and more direct way is to look at the anatomy and physiology of the brain; that is the subject of the next chapter.

8

The brain and perception

Often, the best way to prove a mathematical theorem is to start at both ends, and try to meet in the middle. Starting from the axioms, one proves various propositions, A, B, and C, which look relevant to the theorem. Starting from the theorem, one finds various propositions, X, Y, and Z, such that, if they are true, then the theorem is true. If X, Y, and Z are the same as A, B, and C, then one is home. This principle applies to understanding the brain, in two quite different senses. First, it applies to the ways in which we study the brain. In the last chapter, I discussed the kinds of things the brain must be doing if it is to generate observed behaviour. In the first part of this chapter, I discuss the anatomy and physiology of brains. It would be nice if the two approaches met, but I fear this is true only in the simplest cases. I shall discuss vision, because it is in this field that the two methods are closest to meeting. There is, however, a second sense in which the idea of tackling a problem from both ends, and meeting in the middle, is relevant to the brain. It seems that that is the way the brain itself has to work. In analysing the information that comes in through the eyes, the brain works 'bottom up', piecing together the information from individual retinal cells into larger wholes, and 'top down', comparing pre-existing models in the brain with the visual input. In the second part of this chapter, I try to explain what this means.

Brains are made of cells called neurones. From the cell body of a neurone there emerge two kinds of processes: relatively short dendrites which carry messages towards the cell body, and a single axon, which may be a metre in length, although in the brain it is usually much shorter, which carries signals away from the cell body. Neurones are of three kinds: in sensory neurones the dendrites make contact with a sensory cell, such as a hair cell in the ear or a light-sensitive cell in the retina; in motor neurones, the axons make contact with a muscle fibre; finally, most neurones in the brain make contact only with other neurones. This classification is over-simplified, but it will have to do.

The 'signal' which travels from the cell body out along the axon is an electrical nerve impulse. The nature of this impulse is rather well understood. For our purposes, what matters about it is that it is an all-or-none signal, with an amplitude that does not attenuate as it travels along the axon. The rate of propagation varies from one to a hundred metres per second, depending on the diameter of the axon. A neurone will emit impulses at a rate of from zero up to several hundred per second. Its firing rate depends on the stimuli reaching its dendrites, from sensory cells or other neurones. It is important to appreciate that all a neurone can do is to transmit impulses at varying frequencies. The significance of the message depends on where it comes from. Thus if neurones in the retina of the eye are stimulated electrically, this will be interpreted in the brain as light from a particular direction. In consequence, a blow on the eye causes you to see flashes of light.

Neurones connect with one another via 'synapses', where an end branch of an axon comes into contact with a cell body or dendrite of another neurone. Messages are transmitted chemically: small packets, or 'quanta', of specific 'neurotransmitter' chemicals are released at the synapse, and alter the state of the receiving cell. This alteration may be either excitatory or inhibitory: that is, it may make the receiving cell more or less likely to emit impulses itself. In effect, the cell body integrates up the plus and minus stimuli it is receiving, and this determines its own output. A number of different neurotransmitter substances are used in brains. The significance of this is still far from clear, but it is proving to be of considerable medical importance, because some types of mental illness are associated with abnormal quantities of these substances, and because many drugs with effects on the brain (anaesthetics, tranquillizers, hallucinogens) are chemically related to natural neurotransmitters.

The number of neurones in a brain varies from a few hundred in simple invertebrates up to approximately 10^{11} in man. In many invertebrates, the number and arragement of the neurones is constant for a given species. One of the best known is a marine snail, *Aplysia*, whose study was pioneered by the American physiologist E. R. Kandel. In the brain of this animal, there is not only a small fixed number of cells, but there is also constancy of many other features: the connections made by each cell to others, whether each connection is excitatory or inhibitory, the neuro-transmitter substances released, and whether a cell is spon-taneously active or not. This simplicity has made it possible to work out how particular aspects of behaviour are controlled. For example, the muscles of the heart are spontaneously active, causing rhythmical contractions. There are two neurones which accelerate the heartbeat, and two which slow it down. There are also neurones which cause constriction of the walls of the blood-vessels. Interestingly, there is a single 'command' neurone which can cause an increased supply of blood to the body, by stimulating the two accelerating neurones, and inhibiting the neurones which slow down the heart, and also those that constrict the blood-vessels. This may be a simplified version of the mechanism underlying the fixed action patterns of the ethologists (p. 77). For example, activation of a single neurone in the goldfish causes the animal to flee from danger.

These fixed connections in the brain might suggest that learning is impossible, but this turns out not to be the case. *Aplysia* has a simple reflex in which the gill is withdrawn if the siphon (a fleshy spout leading to the gill chamber) is stimulated. This reflex is brought about by a very simple neural mechanism. Touching the siphon stimulates 24 sensory cells. These synapse with six motor cells, whose axons cause the muscles which withdraw the gill to contract. This reflex is subject to two kinds of learning, called habituation and sensitization. In habituation, repeated stimu-lation of the siphon leads to a gradual weakening of the response. This happens because of changes at the synapses between motor and sensory cells. With repetition, these synapses become less effective in causing the motor neurones to fire. Something is known of the physical nature of the change in the synapse. Repeated training can cause changes that last for several weeks.

In sensitization, stimulation of the siphon is combined with a noxious stimulus to the head. This has the effect of greatly increasing the strength of the withdrawal reflex in response to subsequence stimulation of the siphon alone. This is brought about by a third type of neurone, which forms synapses close to the synaptic terminals of the sensory cells. Activation of this third type of neurone has the effect of increasing the amount of neurotransmitter released at the synapse between sensory and motor neurones, for each nerve impulse travelling down the axon of the sensory cell. Hence it increases the effectiveness of the synapse.

In these experiments, two kinds of learning have been shown to be possible in a nervous system with fixed cell numbers and connections, by altering the effectiveness of synapses. We do not know whether all learning depends on similar processes, or whether, in higher animals, it also involves the creation or loss of connections.

Studies of simple brains like that of *Aplysia* tell us quite a lot. They show the anatomical basis of an unconditioned reflex, the gill withdrawal reflex, and suggest the mechanism of a fixed action pattern. They show that learning can result from changes in the effectiveness of synapses. But they do not tell us much about how more complex cognitive processes take place. The human brain shows a degree of complexity of a different order. There are some 10^{11} neurones, each of which makes some 1,000 synapses, so there are perhaps 10^{14} synapses. It seems reasonably certain that different human brains are not anatomically identical, as different *Aplysia* brains are, although there are striking similarities in the kinds of neurones present in particular regions of the brain, and in their connections.

It is hard to work out the anatomy of the brain. As I said earlier, at first sight it looks like a grey mush. Even under the microscope, it is hard to detect any structure, because of the bewildering variety of nerve fibres criss-crossing one another. Although this book is about ideas rather than techniques, it is important to remember that progress in biology is often made possible by technical advances. In understanding brains, some of the crucial steps have been technical. Anatomical studies have depended on various ways of making visible one cell body and all its processes, while leaving other cells unstained. Physiological studies have depended on the possibility of recording nerve impulses in single cells in the working brain.

In the human brain, the most conspicuous structure, overlying the rest, is the cerebral cortex, a folded sheet of tissue, some one and a half square feet in area if spread flat, containing layers of neurones and their processes. If only because of its sheer size, it is plausible that the cortex is important as a physical basis for higher mental functions. Certainly, it plays a more important role in higher mammals than in lower ones. A mouse without its cortex is not strikingly different in behaviour from a normal mouse: a man with no cortex is a vegetable. There is clear evidence of localization of function in the cortex. In man, this comes in part from studying the effects of damage to specific regions of the brain, for example as a result of a stroke, and in part from stimulating particular regions of the brain during surgery. For example, it has been found that there are two specific regions of the left hemisphere which are necessary for the proper use of language. There is also a region concerned in the recognition of faces. A male patient with damage to this region was unable to recognize his wife or other members of his family by sight: this was not the result of a generalized loss of the ability to recognize people, because he had no difficulty in recognizing them by their voices. Despite this localization, however, there is some flexibility: if one region is damaged, its functions can in part be taken over by others.

A general feature of the cortex is that it contains mappings of the body, and of the world. Thus there is a sensory map of the body, so that stimulation of the skin surface causes electrical activity in the corresponding place in the cortex. The mapping is topologically consistent, so that neighbouring regions of the body map to neighbouring regions of the cortex, but it is distorted in shape, so that regions rich in sensory endings (the lips, the thumb and fingertips) are represented by relatively large cortical regions. There is also a motor map of the body, so that electrical stimulation of the cortex causes movement of the corresponding part of the body.

The part of the cortex whose working we understand best is the primary visual cortex (also known as the striate cortex, or area 17). This is the first region of the cortex to which visual information is transferred. However, some calculations on the visual input have already been performed in the retina, before it is passed to area 17. Thus it is not the case that each retinal cell connects to one cell, and one cell only, in the visual cortex. Indeed, there would not be much sense in such an arrangement. It would, it is true, convey the necessary information from the eye to the brain, but all the necessary computations would still remain to be done. Instead, some of the simpler computations are performed in the retina itself. There are at least two layers of neurones in the retina, in addition to the light-sensitive cells themselves. A given sensory cell is connected to many axons in the optic nerve, and each axon in the optic nerve receives inputs from many sensory cells. The results of these cross-connections can be discovered by recording the impulses in a single axon in the optic nerve, while illuminating the retina in different ways. By these means, D. H. Hubel and T. N. Wiesel were able to discover what optical stimuli were most effective in inducing activity in particular axons. In higher mammals, it turns out that, for most axons in the optic nerve, the most effective optical stimulus is a light spot surrounded by a dark surround. Remembering that impulses arriving at a synapse can be either stimulatory or inhibitory, it is easy to imagine wiring diagrams that would have this effect.

Before going on to ask what happens in area 17, it is worth digressing to point out that things are rather different in other vertebrates. In frogs, for example, stationary illumination of the retina is rather ineffective in causing activity in the optic nerve. By and large, the retina only sends messages to the brain when the image on it changes. This is quite a common feature of sensory cells, including tactile ones. As Hubel remarked 'no one needs or wants to be reminded sixteen hours a day that his shoes are on'. The most striking feature of the frog retina is the presence of 'bug-detectors': that is, fibres in the optic nerve which are stimulated by small dark moving objects on the retina.

Returning to higher mammals, you will recall that fibres entering the primary visual cortex are best activated by spots of light with dark surrounds on the retina. Hubel and Wiesel went on to record output from cortical cells, while illuminating the retina in various ways. They found that the most effective stimuli were oriented lines of various kinds. For different cells, the best stimuli were a line of light with a dark surround, or vice versa, or a boundary between light and dark. The orientation was critical: rotating the line of illumination through as little as ten degrees substantially reduced its effectiveness. At a given point on the cortex, different cells were sensitive to different kinds of lines, with different orientations, but all from the same place on the retina: in other words, there is a spatial map of the visual field in the brain.

Matters are further complicated by the fact that information from the two eyes is brought together in the primary visual cortex. All information from the right visual field, whether entering by the right or left eye, is passed to the left visual cortex, and all information from the left visual field is passed to the right visual cortex. Thus the cortex on one side contains a map of only half the visual field, but it contains two versions of that map, one from each eye, superimposed on one another. Since the images are slightly different, particularly for nearby objects, a comparison of the two versions can give information about distance.

From this brief account, it will be apparent that the retina and primary visual cortex are performing computations on the visual input. In particular, they are extracting local features of the image on the retina—moving bugs in the frog, oriented lines in mammals. But it is a long way from the recognition of such local features to the interpretation of whole scenes. It is one thing to recognize a small dark moving spot, and quite another to recognize that one's colleague Dr X is in a bad temper, or that there is a car approaching on the wrong side of the road. It is therefore important to emphasize that the primary visual cortex is not the end of the road as far as the analysis of visual input is concerned. Fibres from this region pass to other cortical regions. It would be very puzzling if this were not so, because the primary visual cortex simply does not have the structure required to recognize large objects. Its calculations are all relatively local ones: neurones do not make contact with other neurones at any great distance in the same cortical region. Like the retina, it must pass its information on somewhere else for further analysis.[1]

How far does what we know about neurones and about the brain help to explain how we see? In answering this question, I am following the ideas of David Marr. When I say that I can see, I mean that I can identify the three-dimensional shapes of objects, and their relative positions: thus, I can see that there is a chair standing in front of a table with a pile of books on it. To do this, I must rely on the truth of a number of facts about the world, and about the optical properties of my eyes. For example:

i) Objects are bounded by surfaces.

ii) A surface is represented on the retina as a region whose brightness and surface markings change only gradually. (Notice that this is not always true. A shadow may lie across a surface, giving a sharp change of brightness; such exceptions to the rules make the interpretation of images more difficult.) Gradual changes give clues as to the shape and orientation of a surface; for example, if you know where the light is coming from, changes in brightness will tell you whether the surface is concave or convex.

iii) If one surface, A, is closer than another, B, then the image of A relative to B on the retina of the right eye is slightly to the left of the image on the left retina.

These are merely examples of the kinds of facts that make interpretation possible. The first task is to identify surfaces. To do this, it is useful to identify edges, and surface markings. We have already seen that there are cells in the visual cortex that recognize edges. In Figure 11 I show the kind of wiring diagram that would make a single cell responsive to a particular kind of edge, with a particular orientation. What about surface markings?

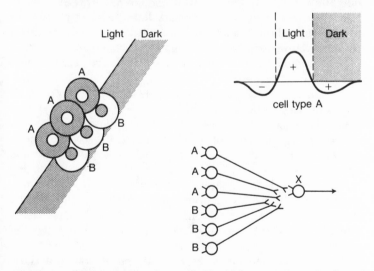

Figure 11. Wiring diagram for a cell, X, which responds to an oriented boundary between light and dark, as shown.

Cell X is stimulated to fire by two kinds of cells, A and B, each of which receives input from a number of retinal cells. Cell type A is stimulated by light at the centre of its field and inhibited by light at the periphery: cell B is inhibited by light at the centre, and stimulated by light at the periphery. If the cells lie as shown relative to the boundary, the net effect is to cause X to fire.

The first job is to recognize particular kinds of marks; for example, the bricks in a wall, the hairs on a cat, the leaves on a tree. With the enormous number of cells present, it is not difficult to imagine interconnections which would make particular cells responsive to particular marks. The next, and perhaps more difficult, stage is to recognize that marks of a particular kind form a pattern over a larger region. It is therefore interesting that there is evidence that the brain is remarkably good at recognizing patterns of similar marks. The experiments, due to Leon Glass, are so simple that they can be repeated by anyone with access to a photo-copying machine. Make a random pattern of, say, square dots on a transparent sheet of paper, and make an identical copy. Lay the two sheets over one another so that the dots coincide, and then rotate one copy slightly. It will be at once obvious that a rotation has occurred, round an identifiable centre. If the degree of rotation was very small, so that for each dot the nearest neighbour is its own image, this would not be surprising: all that need be done is to draw imaginary lines between all pairs of nearest neighbours, and those lines will circle the centre of rotation. However, you will recognize the rotation even when it is large enough to ensure that, in most cases, the nearest neighbour is not the image.

Two further experiments demonstrate that this ability depends on identifying patterns of similar marks. If, in the copy, each square is replaced by some other mark—a hollow triangle, say—you will no longer see a rotation. Suppose, however, you stick to square marks, and make a slightly expanded copy. If this is superimposed, the joint pattern will be seen as lines radiating from a centre. If three patterns—the original, one rotated, and one enlarged—are all superimposed, then no pattern will be detected. But if, say, the original and the rotated dots are pale, and the expanded dots are dark, you will see the pattern of rotation.

These experiments may seem trivial, but they are not. They show that we can see patterns of similar markings. It is not difficult to program a computer to do the same. The significance of this ability is that it helps to identify the surfaces of things. Neurophysiologically, however, the ability to recognize that a particular pattern exists over a sizeable region of the retina cannot be achieved by the primary visual cortex, where neural connections are only local. The job must be done elsewhere.

To see objects, it is not enough to see surfaces, although that is an excellent start. From a knowledge of what edges and surfaces are present, and how they are oriented relative to the eye, we are able to deduce with some confidence what three-dimensional objects are present. How is this done?

There is now a general agreement that to perceive a three-dimensional object requires that one starts out with a set of models in one's head of the kinds of things that might exist, and of what a 2-D image of them might look like, and that one should then test the actual images on one's retina against these models. The first person clearly to express this idea was the German physiologist Hermann von Helmholtz, who wrote in 1866 that to see things is to form 'unconscious conclusions from analogy': by an analogy he meant a pre-existing theory, or model, of what the world is like. It is only now that the correctness of this approach is becoming apparent. I now present three arguments to persuade you that it is indeed correct. I will argue, first, that we are inveterate formers of hypotheses about what we are seeing; second, that computers can only be programmed to see things if they are provided with suitable models; and third, that our habit of guessing what we are seeing is astonishingly successful. But before I do this, let me make a philosophical point. When I first met these ideas, I was reluctant to accept them, because they seemed to suggest that our knowledge of the world is more unreliable than I would like to think. The point is, I think, that the reliability of particular perceptions is indeed low. If we can be reasonably confident about what the world is like, it is because our different perceptions, and the perceptions of different people, are consistent, and, most important, because if we test our perceptions they are usually confirmed: if I see a pile of books on a table, then there are tangible books when I walk over to them.

First, to show that we are incorrigible theorizers. If, in a Rorschach test, I show you a symmetrical ink blot, you do not see merely an ink blot, but a bat, a witch, a face, or a dragon. So deeply ingrained is our instinct to search for a pattern that we refuse to accept any input as genuinely random. One of the oddest features of human history is the way in which we use randomizing devices—the fall of dice, cards, or millet sticks, the tea-leaves in the bottom of a cup, the cracks in a scapula thrown on the fire—to foretell the future. We assume that every pattern has a meaning. This refusal to accept any input as random is the outcome of millions of years of evolution.[2] Most input is meaningful, but the only way to find the meaning is to guess what it might be, and see if the guess makes sense. In treating the fall of dice in this way, we are only doing consciously, and slowly, what we do all the time unconsciously, and almost instantaneously.

The need to build hypotheses into pattern-recognizing programs can best be illustrated by referring to the many programs that have been written to interpret line diagrams (see Figure 12). In effect, these programs assume that the task of extracting edges from the visual image has been completed, and concentrate on converting a set of lines into a 3-D interpretation. All successful programs have one feature in common. They contain a set of hypotheses about what particular features in the 2-D image might be, and a set of rules governing how those features can fit together to form solid objects. For example, consider feature X in Figure 12a. Three lines come together at a point. One possible hypothesis—in fact the correct one—is that three surfaces meet at a point projecting towards the observer. An alternative hypothesis, which would be correct of the identical feature X in Figure 12b, is that the corner projects away from the observer. To decide which of these alternatives is correct in any particular case, the program must be provided with rules governing the ways in which local features can fit together to form solid objects. Thus the interpretation of X as a projecting corner in Figure 12a is made because that interpretation fits together with allowable interpretations of the other features in

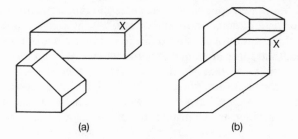

(a)   (b)

Figure 12. Two line-drawings which we perceive as solid objects.
(a) is a drawing which is correctly interpreted by a computer program. (b) is a drawing of the same two objects in different relative positions. Feature X has a different interpretation in the two drawings.

the picture, to form a set of possible solid objects; the same is true of the different interpretation of X in Figure 12b. Of course, there can be pictures that are ambiguous, in that they can be interpreted in more than one way, and even pictures for which there is no plausible interpretation (Figure 13).

A picture-recognizing computer must be provided with the necessary set of hypotheses and rules. Animals may also be born with some such rules: for example, a herring gull chick is born with the hypothesis that a thin projection with a spot on it is a parent's beak (see p. 77). But most of the models in our heads have been learnt. Sometimes, they enable us to extract a lot of information from very few cues. In a remarkable demonstration of this, Johansson and Maas took cine-films of human figures with lights attached to their joints—knees, elbows, and so on—but otherwise in the dark. If the figures are moving, they are perceived as what they are. To do this, we must have in our heads a precise but flexible model of how people move.

In this chapter, I have indicated the kinds of things we know about brains, and the kinds of questions we have to ask if we are to understand how we perceive things. To conclude, I will list what these questions are, and remind you of some of the answers that can be given:

i) What information is in principle present in the image on the retina which could be used to decide what objects are present? Examples were given concerning the images of surfaces.

ii) What calculations could be performed to extract this information? An example of such a calculation would be one which compared the images on the two retinas, and searched for differences which would provide information about relative distances. Another example would be a calculation which

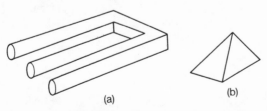

Figure 13. Two line-drawings.
(a) has no consistent 3-D interpretation. (b) is ambiguous: for example, it could be a pyramid, or a rectangular piece of paper folded along a diagonal.

detected the existence of a set of similar markings over some region of the image.

iii) Can psychological experiments give us a clue to how such calculations are performed? One example, suggesting that we may be able mentally to rotate objects, was mentioned at the end of the last chapter, and a second, in this one, demonstrated that we do in fact recognize sets of similar markings (note that a psychological demonstration that we can do something, and an account of a calculation which would enable us to do so, are quite different things).

iv) How is information about the surfaces and edges present in an image converted into a 3-D perception? We can approach this question by writing computer programs to do a similar job, or by finding out what is the minimum information we need in order to perceive something.

v) Physically, how are the calculations performed? This is the domain of the neurophysiologist. Relevant findings include the existence of excitatory and inhibitory connections between neurones, the changes in synapses that occur with learning, and the identification of single cells in the primary visual cortex which respond to edges of particular kinds, with particular orientations, at particular points on the retina.

It is perhaps natural and inevitable that scientists adopting one or other of these approaches should become convinced that their own approach is the best. I suspect that all will be needed.

9

Development

There are two ways we can view the structure of an animal. Consider, for example, the eye spots on the front wings of a peacock butterfly. One approach is to seek an adaptive explanation. We can suggest that the eye spots scare off predators, and there is some evidence that this is so. But even if this adaptive explanation is correct, it does not tell us how the eye spots are made. Indeed, understanding how structures develop is one of the major problem areas in biology. One reason why we find it so hard to understand the development of form may be that we do not make machines that develop: often, we understand biological phenomena only when we have invented machines with similar properties. The shapes of the things we make are, as a general rule, imposed upon them from the outside: we do not make 'embryo' machines which acquire complex shapes by intrinsic processes.

How does a geometrically simple egg turn into a complex adult? Historically, there have been two rival views. Preformationists held that there must be, in the egg, a miniature but perfectly formed adult: 'development' then consists merely of growth in size. This view avoids the necessity of explaining how an increase in complexity can occur by denying that it happens, but only at the expense of supposing not only that there is a minute homunculus in the egg but that within that homunculus there is an egg containing a still more minute homunculus, and so on, in Chinese box fashion, *ad infinitum*—or, if not *ad infinitum*, at least back to Eve, who carried within her a sufficient number of successively smaller homunculi to account for all the future generations of mankind. The preformationist view may just be tenable by a believer in the Garden of Eden: an evolutionary biologist must reject it, and accept the alternative 'epigenetic' view, that there is a genuine increase in spatial complexity during development.

It is popular nowadays to say that morphogenesis (that is, the development of form) is programmed by the genes. I think that this statement, although in a sense true, is unhelpful. Unless we understand how the program works, the statement gives us a false impression that we understand something when we do not. Thus we understand rather well how the genetic message, the sequence of bases in DNA, is translated into a series of proteins. However, a small bag of water containing all the different kinds of proteins—perhaps 10,000 to 100,000 kinds—that the genes of an elephant can code for is not an elephant, and would not turn into an elephant.

How far, then, does our knowledge of how genes code for proteins enable us to explain the shapes of organisms? It does take us some of the way, but not all the way. I explained in Chapter 2 how the sequence of bases in DNA is translated into the sequence of amino acids in proteins. The finished string of a few hundred amino acids will, in a suitable liquid medium at a suitable temperature, fold up to form a three-dimensional globular structure, the active protein. Sometimes, different protein molecules will fit together, like the pieces of a three-dimensional jigsaw, to form larger-scale structures. In effect, the pieces are jiggled about until the appropriate interlocking surfaces come into contact, and then they stick. This process, called 'self-assembly', can explain the formation of some intracellular structures, such as the ribosomes described on page 16. These organelles contain some 50 different kinds of proteins and RNA molecules. It is almost true to say that we know how the genetic program determines the shape of a ribosome. The DNA base sequence determines the amino acid sequence of the proteins: the amino acid sequences determine the 3-D shapes of the proteins: the 3-D proteins fit together, like the pieces of a jigsaw, to form a ribosome.

It is tempting to suggest that the same kind of jigsaw process will explain the shape of an elephant, but almost certainly it is not true. The skulls of an elephant and of a man are different shapes, but this is not because the molecules which fit together to make an elephant's skull are a different shape from those that make a human skull. The shape of a jigsaw is determined by the shapes of the pieces, but there are objects with well-defined shapes which are not explicable by the shapes of their constituent molecules. For example, a breaking wave has a form which is not random, because it is repeated as each successive wave breaks, but the shape is in no sense deducible from the shape of a water molecule: if the sea was made of alcohol, the waves would be much the same shape.

In one sense, then, the shape of an elephant, unlike that of a ribosome, is independent of the shapes of its constituent molecules. Yet in another sense, there are specific molecules that do determine the shape of an elephant, because we know that, if we altered a gene in the fertilized egg, we might alter the shape of the future elephant.

One way of understanding structures larger than ribosomes is to remember that organisms are made of different kinds of cells—muscle cells, bone cells, nerve cells, and so on—arranged in space. How do cells become different? When I learnt biology, I was taught that Weismann, although he was right about the independence of germ line and soma, had given the wrong answer to this question. He argued that cells become different because, at the time of cell division, different genes are transmitted to different daughter cells (he did not use the word gene, but that is what he meant); whereas in fact all body cells contain the same genes, and become different because different genes are activated in different cells. Now it is true that Weismann was mistaken in thinking that different cells receive different genes, but his reasons for thinking so are illuminating.

He saw that the alternative was to suppose that cells become different because they are exposed to different external influences. He knew that this sometimes happens. For example, if a newt's leg is amputated, it will grow a new one. When this happens, cells do things they would not otherwise have done, presumably because, after amputation, they find themselves in a changed environment. A clearer proof that cells change when exposed to new influences from outside is afforded by embryonic induction, first described by Spemann in 1924. As an example of induction, consider the development of the vertebrate eye. The eye first appears as a cup-shaped outgrowth from the brain. The cup becomes the retina, and the stalk connecting it to the brain the optic nerve. As the cup grows out, it comes into contact with the outer cell layer of the head of the embryo. These outer cells, which would otherwise become ordinary skin cells, then differentiate as the lens of the eye: it is the contact with the eye cup that induces them to do so.

In such cases, Weismann speaks of the 'determinants' (that is, genes) being 'liberated' by some specific stimulus. Although he knew that this happens, he did not think it plausible that all differences between cells could be caused in this way, because he could not see how a sufficient number of specific external stimuli could be brought to bear during development. It was for this reason that he concluded that different cells receive different genes. He was wrong, but it is only fair to add that we still do not know how all the specific external stimuli which induce cell differences arise.

I have said several times that Weismann was wrong. How do we know? There are several lines of evidence. The simplest is that, if we look at cell division, we see that identical sets of chromosomes are being passed to both daughter cells. More conclusive evidence comes from experiments in which the nucleus is removed from a fertilized egg, and is replaced by the nucleus of a cell which is already set on the road to becoming, for example, a gut cell or a skin cell. If such cells develop as they do because they contain only some of the genes—for example, just the genes for being a gut cell—then a nucleus from such a cell could not substitute for an egg nucleus. It turns out, however, that an egg with a replacement nucleus can, at least sometimes, develop normally. This shows that it contains all the genes, not just some of them. More recently, this conclusion has been confirmed by the more direct method of extracting DNA from various kinds of cells and determining the base sequence of some of their genes.

We are therefore brought back to the problem which Weismann felt was insoluble. How can a sufficient number of specific stimuli arise during development to elicit different activities from cells? How does it happen that these stimuli are so arranged in space that the right kinds of cells appear in the right places? I have already given one example of how this can happen, in my description of the induction of a lens by the eye cup. The lens forms in the right place, above the eye cup, because it is induced to do so by contact. In animals, the formation of sheets of cells which fold in various ways, and which in consequence come into contact with each other at specific points, is a common feature of development, and provides a partial answer to Weismann's question.

A second cause of differentiation is illustrated by the way in which 'primordial germ cells' are formed in the fruitfly *Drosophila*. There is a special kind of cytoplasm, called the pole plasm, already present at one end of the fertilized egg. As the egg is divided up into a large number of cells, a group of cells at one end come to contain this special pole plasm. It is these cells which will later give rise, by further cell division, to eggs or sperm. If for any reason no cells are formed containing pole plasm—and this does sometimes happen—the resulting flies are sterile for life.[1] This is not the only case in which it is known that the egg contains a special localized material which causes cells containing it to follow a particular path of differentiation. However, the number of differences generated in this way is certainly small.

The localization of material in the egg accounts for some of the differences between cells, and more are explained by induction, when folded sheets of cells come into contact. There are, however, many cases in which differentiation occurs within what at first looks like a homogeneous sheet of tissue. Sometimes it can be shown that the tissue not only looks homogeneous, but actually is so, because, if the tissue is cut in half, each half will regulate to give rise to the same pattern as would otherwise have been produced by the whole. How are we to explain these cases of pattern generation within homogeneous tissues?

I think that the decisive step towards an answer was taken by the English mathematician Alan Turing. To most people, Turing is best known for his work on the theory of computing machines, but in time his work on the theory of morphogenesis may be seen as equally important. What he did was to show that a belief which seems only common sense is in fact false. The belief in question is that diffusion will cause things to become all mixed up: that it will destroy spatial differentiation. Certainly, this is usually true. If you put a drop of soluble dye in a glass of water, it will slowly diffuse outwards until the water is uniformly coloured. What Turing showed is that this is not a universal truth. To be more specific, he showed that, if chemical substances react in a medium through which they can diffuse, it is not always true that the reactants will become uniformly distributed. For particular kinds of chemical reaction, and for particular diffusion rates, it turns out that, even if the reactants are at first distributed uniformly throughout the medium, a regular pattern appears, with substances concentrated in some regions and almost absent from others.

This is a contra-intuitive result. Since Turing's paper, however, an actual chemical reaction, Zhabotinsky's reaction, that generates a pattern, has been described. The actual reacting substances need not concern us. The relevant fact is that, if two solutions are mixed in a petri dish or other container, and allowed to stand, the mixture is at first a uniform brown, but this homogeneous state is not stable. At some point in the dish, a blue spot will appear, and will grow into a ring: then a new blue spot will appear at the centre, and then another, until a set of expanding concentric rings appears. Other patterns are also possible. Ultimately, when the chemical reaction is completed, the pattern is destroyed by diffusion. Thus the pattern is a 'dissipative structure' of the kind discussed on p. 2: it is a pattern maintained by a continuing input of energy, in this case from the chemical reaction.

Turing's model, and Zhabotinsky's reaction, show how wave-like patterns can develop, with regularly spaced regions of high and low concentration. I am tempted to think that some such process is involved in the development of repeated structures, such as the segments of a worm, or a flower's petals, or the stripes on a zebra. Even simpler would be a gradient of concentration, from high to low. Such a gradient would arise if a substance were synthesized at one end, or along one edge, of a region, and destroyed at the other. Of course, it is not enough to show that a gradient of some kind exists: for example, that a frog's egg is darker at one end than at the other. We have also to show that the gradient influences subsequent development. One of the simplest and most elegant demonstrations of this was by Peter Lawrence, as part of the work he did for a Ph.D at Cambridge. He studied the pattern of small bristles on the abdomen of a bug, *Onchopeltus* (see Figure 14). On each segment of the abdomen, the bristles are uniformly distributed, and all point posteriorly. The pattern is so simple that it hardly seems worth studying. If, however, part of the intersegmental membrane separating two segments is missing, something very surprising happens (Figure 14b). The bristles are now arranged in two vortices: why should this be?

Lawrence's explanation is also shown in the figure. He supposes

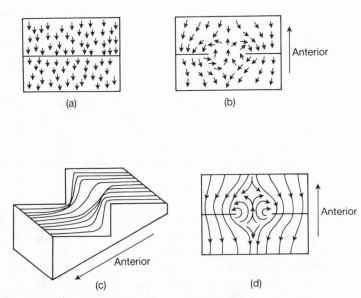

Figure 14. The pattern of bristles on a bug (after Peter Lawrence).
(a) shows two abdominal segments, with arrows indicating the direction in which the bristles point. (b) shows the bristles in an atypical individual, in which the central part of the intersegmental membrane is missing. (c) shows the concentration of a hypothetical substance produced at the front edge of a segment and destroyed at the back edge; the substance has diffused forwards in the central region, where the membrane is missing. (d) is derived from (c), and shows arrows pointing down the concentration gradient.

that, in each abdominal segment, there is a gradient of some diffusible substance, from high at the front to low at the back. The intersegmental membrane does not permit diffusion, so that a high concentration at the front end of one segment is maintained close to a low concentration at the back of the segment in front. Lawrence then supposes that bristles grow so as to point down the concentration gradient. The result is the simple pattern of Figure 14a. What happens if part of the intersegmental membrane is missing? The substance will then diffuse from the posterior segment into the anterior one, giving concentrations as shown in Figure 14c. If bristles then grow so as to point down the local gradient, they will produce exactly the pair of vortices that were observed (Figure 14d).

I find this argument convincing: others may not. Lawrence did not identify any substance which was distributed as his model requires. Indeed, he could modify his model by supposing that the gradient has its high point at the back edge of each segment, and that bristles point up the slope, and it would work just as well. People who only believe in things when they know what they are made of will feel uneasy. My own view is that scientific theories usually start out by assuming the existence of entities that no one has ever seen or touched—genes, atoms, photons, viruses. If the theories are successful, someone will find a more direct way of showing that the hypothetical entities are actually there.

The processes that generate patterns during development are almost certainly more complex than Zhabotinsky's reaction, or the process imagined by Turing, or the generation of a gradient by synthesizing a substance at one point and destroying it at another. Electrical phenomena, and perhaps mechanical ones, may be involved as well as, or instead of, chemical reaction and diffusion. The important point is that, if energy is supplied, a homogeneous spatial field can become inhomogeneous. If, as a result, a particular chemical substance is concentrated at particular places, it may induce the cells in those places to differentiate in particular ways, just as the presence of galactose in the medium can induce bacterial cells to produce proteins they would not otherwise make. There is the difference, of course, that the changes induced in the cells of higher organisms are far more stable and long-lasting than those induced in bacteria.

It would be a mistake to imagine that, in development, the full complexity of the adult structure is generated in a single pattern-forming process. Development occurs in a series of stages, the conclusion of each stage being the starting-point of the next. To give a concrete example, there is in the early development of vertebrates a process called gastrulation, during which a hollow ball of cells, the blastula, is folded in on itself to form a ball of two cell layers, with an opening, the blastopore, at one end. When gastrulation is completed, the cells can be divided into a small number of classes according to their future fate: they may be 'ectodermal' (destined to give rise to the outer layers of the skin, hair, the lens of the eye, and so on), 'endodermal' (destined to give rise to the lining of the gut), 'mesodermal' (destined to give rise to muscle, bone, blood-vessels, and many other structures), or, finally, they may be destined to give rise to the brain and nervous system. These categories are fixed: one cannot readily induce an ectodermal cell to turn into muscle or bone. But within these categories, each cell has a wide range of potentialities: we have already seen that an ectodermal cell can give rise to skin or, if induced by the eye cup, to lens. These potentialities are successively narrowed down in subsequent stages of development.

We can now ask how far it is sensible to say that genes 'program' or 'control' development. Imagine a Turing-like process generating a pattern in a sheet of cells, and then suppose that the local concentration of some substance activates particular sets of genes in particular places. Genes will be involved in this process in two ways. First, the chemical reactions responsible for the pattern will be catalysed by enzymes, which in turn will be specified by genes. Consequently, a change in a gene could destroy the pattern, or alter its form, merely by altering the rate at which some reaction proceeds. Second, changes in genes can alter the responses of cells locally to the inducing chemical: different genes could be switched on by the same stimulus.

The American geneticist Curt Stern was the first to distinguish between these two ways in which a change in a gene can alter the resulting structure: he referred to them as changes in 'prepattern' and in 'competence'. The observations that led him to make the distinction are deceptively simple. He developed a technique whereby one can produce fruitflies (*Drosophila*) in which patches of tissue are genetically different from the rest of the body, usually by having two doses of some recessive mutant gene which is present only in a single dose in the rest of the fly. Figure 15 shows a particular example. There is a recessive mutant, achaete, in which a particular pair of large bristles are missing. Stern examined flies which were genetically wild-type, but had a small patch of achaete tissue at the site where the bristle should be, and achaete flies with a small patch of wild-type tissue at the site. These are shown in the figure.

The interpretation of these results is as follows. Before the development of the bristles themselves, a prepattern arises in the tissue: I imagine this as a series of peaks of concentration of some inducing chemical, arising by a 'Turing' process, but this is not necessary to Stern's argument. Each peak induces a cell locally to

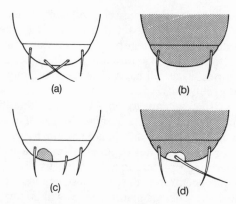

Figure 15. The bristles on the scutellum of *Drosophila*, showing an experiment by Curt Stern.

(a), a normal fly; (b), an achaete fly; (c), a normal fly with a patch of achaete tissue; (d), an achaete fly with a patch of normal tissue. In Stern's experiments, the normal and achaete tissues were made visible by incorporating a second mutation affecting cuticle colour. The essential finding is that a patch of normal tissue on an otherwise achaete fly produces a bristle, whereas a patch of achaete tissue on an otherwise normal fly does not.

divide and form a bristle. In achaete flies, the prepattern forms normally, but the cells locally do not respond, presumably because some gene needed for the response is changed. For most other mutant genes that Stern studied in this way, the abnormality in structure could best be explained by supposing that cells could not respond to the prepattern, rather than that the prepattern itself had changed. It would be wrong, however, to regard prepatterns as immutable and independent of genes. Whether they arise as Turing proposed, or by some other physico-chemical process, it will always be true that they can be altered by changes in genes.

Since Stern's work, studies of the relation between genes and development have become far more sophisticated, but the basic concepts are still those discussed in this chapter. Recent studies rely on the one hand on techniques of transplanting tissues, or cells, or even individual genes, from one site to another, and on the other of identifying the genes concerned in particular processes. Morphogenesis is seen as a combination of pattern-forming processes, and of the activation and inactivation of genes. Turing's model is a prototype of a pattern-forming process which is perhaps too simple, but which does seem to be the kind of process we are looking for. As a model of gene activation, the mechanism discovered by Jacob and Monod in bacteria again provides a satisfactory prototype, although we know that the changes that occur in eukaryotic cells are different at a molecular level, and far more stable.

Our understanding of development is at a stage reminiscent of genetics in the 1930s. At that time, thanks in particular to the work of Morgan and his colleagues on *Drosophila*, we had an abstract model of how genes are arranged on chromosomes, and of how they are transmitted from generation to generation, but we did not understand the chemistry of genes, or their replication, or their role in protein synthesis. Today we are beginning to acquire an abstract knowledge of development, in terms of gradients and prepatterns and gene-switching. But although we may be confident that a gradient exists, and influences subsequent development, we usually have no idea what it is a gradient of, and rather little idea of how genes are actually switched on and off. More precisely, lots of people have ideas, but they do not all agree.

10

The origin of life

Darwin ended *The Origin of Species* with the words:

> There is a grandeur in this view of life, with its several powers, having been originally breathed by the Creator into a few forms or into one; and that, while this planet has gone cycling on according to the fixed law of gravity, from so simple a beginning endless forms most beautiful and most wonderful have been, and are being evolved.

Although he later regretted it, he was willing tó hand over the origin of life to the Creator because, I imagine, he saw no prospect of an immediate experimental attack on the problem. Things are very different today. We have an increasingly clear idea of the various stages that must have occurred in the transition from chemistry to biology. Some of these stages are quite well understood, and all are under active investigation. We are still some way from a full solution, but to achieve such a solution would be a big step towards understanding the problems of biology.

It is best to start with a restatement of the problem. Living entities have the properties of multiplication, variation, and heredity. To understand the origin of life on earth requires two things. First, we must understand how entities with these properties could arise naturally in the conditions existing on earth. Once this had happened, evolution by natural selection would inevitably follow. Since, however, the first living things must have been far simpler than anything living now, a second requirement is to show how these first simple entities could evolve by stages into organisms similar to existing ones.

The plan of this chapter is as follows. I first describe the two earliest stages in the origin of life: the origin of organic molecules, and the origin of polymers—that is, of macromolecules formed by stringing organic compounds together. Both these steps are essentially chemical, and are of necessity discussed rather briefly. I then describe three series of experiments, on 'coacervates', on 'proteinoids' and 'microspheres', and on 'naked genes'—which illustrate how close we are to bridging the gap. Yet each of these experiments fails, in crucial ways, to be a complete solution. They all leave unsolved the central problem of the origin of a hereditary mechanism based on nucleic acids, able to code for all the enzymes needed for their own replication. The last part of the chapter discusses how this problem may be solved.

Serious work on the origin of life started in 1932 with a suggestion, made independently by J. B. S. Haldane and by the Russian biologist A. I. Oparin, that the atmosphere of the primitive earth may have contained no free oxygen. The first step towards life is the chemical generation of 'organic' compounds: that is, compounds containing carbon. Such compounds react very readily with oxygen. In an atmosphere containing oxygen, they would not last very long. The origin of life, therefore, requires the absence of free oxygen. There is some direct evidence that oxygen was indeed absent from the primitive atmosphere. It is absent from the atmosphere of other planets in the solar system. More direct, iron in the presence of oxygen is converted into a red-brown oxide (Fe_2O_3): in the oldest rocks it is found in a less oxidized form.

It is worth digressing for a moment to discuss the source of the oxygen now present in the atmosphere. The short answer is that green plants use the energy in sunlight to split water molecules, using the hydrogen in the synthesis of organic compounds, and releasing the oxygen. Things, however, are not quite so simple. When a plant dies, the chemical decomposition of its tissues uses up exactly the quantity of oxygen it liberated while alive: the life, death, and decomposition of a tree does not add a single molecule of oxygen to the atmosphere. If there is to be a net increase of oxygen, the plant remains must not be oxidized, but must be preserved in non-oxidizing conditions, ultimately becoming coal or oil. When I first realized this, I formed an alarming picture of the last motorist using up the last molecule of oxygen in the atmosphere to burn the last gallon of petrol in his tank. Perhaps fortunately, however, most of the fossil carbon is in places where we cannot easily get at it.[1]

Haldane and Oparin suggested that, in the absence of free oxygen, organic compounds would arise naturally. The suggestion was taken up in Chicago in 1953 by Stanley Miller and Harold Urey. Miller passed electrical discharges (mimicking lightning) through a gas consisting of water vapour, methane, hydrogen, and ammonia, all substances that may well have been present in the primitive atmosphere. The result was the generation of a range of organic compounds, including many of the natural amino acids. Since that time, almost all of the organic compounds of which living matter is composed have been synthesized in the laboratory in similar experiments. On the primitive earth, the ultraviolet light from the sun may have been a more important source of energy than electrical discharges. Today, UV light is absorbed in the upper atmosphere by the ozone layer, but in an atmosphere lacking oxygen there would have been no ozone. In the primitive oceans, with no oxygen to react with them and no microorganisms to feed on them, organic compounds would have accumulated until the water, in Haldane's phrase, became a hot dilute soup.

The appearance of organic compounds, then, is fairly well understood. The next stage is the linking of these compounds to form polymers, in particular proteins and nucleic acids. There is a difficulty here that is easily stated. The polymerization reaction, both for proteins and nucleic acids, involves the removal of a molecule of water. This does not happen readily to substances in solution in water; there must, therefore, have been some way in which the substances were concentrated. To make matters worse, the reaction requires an input of energy. In existing organisms, polymerization is carried out by enzymes, and the energy is supplied by ATP (see p. 62). As it happens, inorganic phosphates (that is, phosphates without the adenine 'handle' of ATP) may well have been present in the primitive oceans; if so, they could have supplied the energy for polymerization.

It is also fairly easy to think of ways in which the reacting substances could have been concentrated. This could happen by the evaporation of water in tidal pools, or by freezing, which removes pure ice and so leaves behind a stronger solution, as applejack is made by removing ice from cider. Perhaps the most promising mechanism for concentrating organic compounds, however, is by absorption on the surface of minerals. Clays, for example, consist of stacked sheets of molecules, between which water and dissolved substances can percolate, so that the effective surface is enormous. Aharon Katchalsky of the Weizmann Institute in Israel has shown that clays will promote the polymerization of amino acids to form protein-like chains, provided that energy is provided in the form of phosphates. The main remaining difficulty in understanding the origin of biological polymers arises from the fact that there are a number of chemically different bonds by which amino acids and nucleotides can be linked together. Only one kind of bond is used in biological polymers, but several different ones are liable to form when attempts are made to mimic events that might have occurred on the primitive earth.

So far, the problems I have discussed are essentially chemical. I now turn to attempts to simulate proto-life in the laboratory. One that admirably reproduces some of the features of living cells is a long series of experiments by Oparin. If various polymers—proteins, nucleic acids or carbohydrates—are dissolved in water, they tend to form droplets, which Oparin has called coacervates. Any other substances present which are more soluble in the material of the droplet than in water will tend naturally to be concentrated within the droplets. In one particular experiment, Oparin studied droplets formed from histone (a protein) and gum arabic (a carbohydrate). If he added an enzyme (derived, of course, from a living cell), able to link sugars to form starch, this enzyme accumulated in the coacervate droplets. If he then added a suitable sugar (glucose), linked to a phosphate to provide energy, the sugar molecules diffused into the droplets, where they were linked together to form starch. The starch remained within the droplets, and the released phosphates diffused out. As a result the droplets grew and split into two; each daughter droplet continued to grow, provided that it, too, contained enzyme molecules. This process of multiplication continued, until the enzyme originally provided was diluted out.

Oparin's coacervate droplets have a metabolism, and they grow and divide. They do so, however, only because they are provided with an enzyme synthesized by a living organism. Further, they lack any mechanism for replicating hereditary information, and so cannot evolve. Sidney Fox of the University of Miami has followed a parallel course. He has found that if mixtures of amino acids are dried and heated to 130° C, they rapidly form polymers, which he calls proteinoids, in which most but not all of the chemical linkages are those characteristic of biological proteins. He proposes that volcanoes would have provided a suitable environment on the primitive earth. If the proteinoids are now dissolved in water, they form small microspheres, bounded by a membrane, which grow and bud off smaller spheres. Fox finds that his microspheres have rather non-specific enzymic activities: that is, they catalyse a range of chemical reactions. Like Oparin, Fox has produced objects that grow and divide, given a suitable environment. Their 'metabolism' is less specific than that of Oparin's coacervates, but then Fox has not added specific biological enzymes. Crucially, however, Fox's microspheres lack heredity, and so will not evolve by natural selection.

I turn now to a series of experiments of a very different kind, in which the aim has been to mimic not growth and metabolism but heredity and evolution. The entities which evolve are simply molecules of RNA. This polymer is chosen, rather than DNA, because it is single-stranded, and consequently folds back on itself to form hairpin and cloverleaf structures (see Figure 16). An RNA molecule therefore has a 'phenotype' which can affect its stability and ease of replication: that is, its Darwinian Fitness. The problem is to provide an environment in which RNA molecules will multiply and evolve. This consists of a test-tube containing a solution of the four bases from which RNA is made, and an enzyme known as a 'replicase', coded for by an RNA virus called $Q\beta$, which infects the bacterium *E. coli*. Given an RNA molecule as a template, and the necessary bases, this enzyme will produce new RNA molecules with the same base sequence (of course, it first produces molecules with the complementary sequence, and from these molecules with the original sequence).

A typical experiment, then, proceeds as follows (see Figure 17). A few similar RNA molecules are added to a test-tube containing bases and enzyme. After some hours, when many copies have been made of each of the original molecules, a drop of the resulting solution is added as a seed to another test-tube containing bases and enzyme. This process is repeated many times, and the nature of the RNA molecules is followed. Such

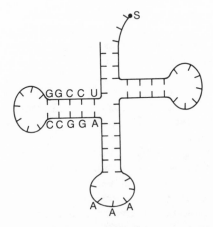

Figure 16. Diagram of a tRNA molecule.

The anticodon AAA is at the bottom. The attachment site, S, for the amino acid is at the top. The base sequence is such that, when the molecule bends back on itself, the appropriate base pairings can be achieved. This is illustrated on the arm projecting to the left; elsewhere the names of the individual bases have been omitted.

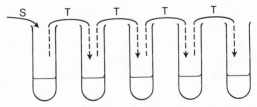

Figure 17. The evolution of RNA molecules.

The test tubes contain an enzyme that will replicate RNA, and the four bases. S indicates that RNA molecules are added as a seed to the first tube, and T that, at intervals, a drop of solution is transferred from one tube to the next.

experiments were first performed by the American Sol Spiegelman, and have been pursued further in Manfred Eigen's laboratory in Göttingen.

Before describing the results, it is important to appreciate that what is being followed is the process of natural selection in an evolving population of molecules. Those kinds of RNA molecule which are replicated particularly rapidly and efficiently will replace all others. New variants will arise in the course of the experiment, because replication is not exact. Sometimes part of a molecule will be replicated twice, or it will not be replicated at all. Also, changes of single bases may occur. In fact, there is a chance of the order of 1 in 10,000 that an inappropriate base will be incorporated as each base is added: this rate is often called the 'error rate'.

In his original experiments, Spiegelman found that the 'winning' molecule is some 220 bases long, even if he started with a much longer molecule. This is perhaps not surprising; if all you have to do is to get yourself replicated, it pays to be short. The reason why the winner was not even shorter is that the enzyme finds it hard to recognize very short molecules. More surprising, when the experiment has been repeated elsewhere, the winner has proved to have a sequence identical or very similar to Spiegelman's molecule. It seems that, in the very simple environment of these test-tubes, there is one unique best kind of molecule to be. The evolution of a complex and varied set of organisms may require a varied physical environment. If the environment in the test-tube is changed, however, then the base sequence of the winning molecule is also changed. For example, if a drug is added which inhibits replication by binding to the RNA and interfering with the enzyme, then molecules evolve which lack suitable binding sites for the drug.

These experiments show that natural selection between molecules can occur, and that it has repeatable results. It is not, however, an adequate model of what went on in the primitive soup, because replication depended on the presence of an enzyme which was a good deal more complex, in terms of the amount of information needed to specify it, than the molecules it was replicating. Nevertheless, it may well be that replicating nucleic acid molecules were the first living things, in the sense that they were the first entities with multiplication, variation, and heredity. For this to be true, it must be possible for nucleic acids to replicate in the absence of 'informed' enzymes: that is, of enzymes themselves coded for by genes. Thus it is unreasonable to imagine that molecules of Q β replicase were present in the primitive oceans, although there may well have been proteinoids of the kind studied by Fox, and other substances with rather non-specific catalytic activity.

Is it possible for nucleic acids to replicate in the absence of enzymes? The work of Leslie Orgel, an Englishman working in

California, suggests that it is. Remember that RNA contains four bases, G, C, A, and U, and that in replication G pairs with C and A with U. In a typical experiment, Orgel started with a solution of A and G monomers and added as a template RNA molecules containing only the base C. The result was that a complementary chain was formed, up to 40 bases long, consisting largely but not entirely of the base G, as expected if the appropriate pairing was occurring. This result does not require the presence of an enzyme, but does depend on the presence of zinc as an inorganic catalyst.

What, then, is the relevance of enzymes for the replication of RNA? Referring back to p. 60 and Figure 8, we saw that enzymes speed up chemical reactions by lowering the activation energy. In the early days of life, speed of replication was not essential: in the absence of competition there was plenty of time. What was much more important was the 'accuracy' of replication: that is, the precision with which G is paired with C and A with U. By lowering the activation energy of the appropriate pairings, and raising that of the inappropriate ones, an enzyme can increase this accuracy. I mentioned earlier that, when RNA is replicated by the Q β replicase, the error rate is approximately 1 in 10,000. The corresponding rate in the absence of the enzyme, in experiments such as Orgel's, is about 1 in 10, and certainly greater than 1 in 100.

Why does this matter? Essentially, because the quantity of genetic information in any organism is limited by the accuracy with which that information is replicated. Thus imagine, for simplicity, a population of asexual organisms, each of which can, on average, have ten offspring before it dies. If the population is to be maintained without steady genetic deterioration, one of these ten must have the same genetic information as its parent. The other nine can have mutations that render them less able to survive: if so, they will be eliminated by natural selection. But if not even one of the offspring is without mutations, the genetic information will gradually decay. Suppose, to take a numerical example, the organism had a genome of 10,000 bases, and that these were replicated with an error rate of 1 in 1,000. The chance that each new base is correct is then 999/1,000, and that all 10,000 bases are correct is $(999/1000)^{10,000}$, or approximately 1 in 22,000: with only ten offspring, there would be little chance that any were identical to their parent. Very approximately, if the population is to survive, and if it has 10,000 bases in its genome, the error rate must not be greater than 1/10,000.

The size of the Q β virus is 4,500 bases, which is close to the upper limit of size for an organism whose replicase has an error rate of 1 in 10,000. Although it is far from the problem of life's origins, it is worth digressing to ask how organisms as complex as ourselves manage to replicate our genomes. In effect, we have a 'proof-reading' stage. Our replicating enzyme first puts in a base, with an error rate of 1 in 10,000, similar to that of Q β replicase, and then checks it, and replaces it if it is wrong. The second stage also has an error rate of 1 in 10,000, so the overall rate is about 1 in 10^8.

This has brought us to the catch-22 of the origin of life. The first replicating molecules had to manage without informed enzymes, and hence had to put up with error rates greater than 1 in 100. This limited their genome sizes to 100 bases or less. To improve on this, they had to code for a replicase enzyme, and also for a primitive protein-synthesizing machinery. That cannot be done with as few as 100 bases. So, if you cannot increase your genome size you cannot code for an enzyme, and if you cannot code for an enzyme you cannot increase your genome.

A possible way out of this impasse has been suggested by Manfred Eigen and Peter Schuster. Imagine that you wish to replicate the message GOD SAVE THE QUEEN. You are allowed to make ten copies, and keep the best one. Your error rate per letter is 1 in 5: this means that only one copy in 28 is perfect, and that is not good enough. What can you do? One suggestion is to replicate and select each word separately. That is, replicate QUEEN ten times, and select the best copy. With an error rate of 1 letter in 5, you should be reasonably certain of getting at least one perfect copy. This would work, but there is a snag. In molecular terms, the suggestion is that a more complex genome could arise by dividing the total information into sections, and imposing natural selection independently on each one. The snag is this: how do you prevent one of the sections out-competing all the others? Thus suppose that the four words represent different molecules, all of which are competing for the same bases. If one replicates faster than the others, it will in time replace them. Your message will soon be reduced to the one word, GOD.

Figure 18 shows the solution suggested by Eigen and Schuster. The four words, or the molecules they represent, are arranged in a 'hypercycle', such that the rate of replication of each molecule in the cycle depends on the concentration of the molecule immediately before it in the cycle: the rate of synthesis of SAVE is proportional to the concentration of GOD, of THE to the concentration of SAVE, of QUEEN to the concentration of THE, and, finally, of GOD to the concentration of QUEEN. If, for example, each molecule acted as a template encouraging the synthesis of the next in the cycle (as mRNA encourages the synthesis of the corresponding protein—see p. 17), then the required relationships would exist. It can be shown mathematically that if these relationships do exist, then the whole cycle is stable: no one molecule replaces all the rest. Intuitively, the reason is that, if the concentration of any molecule rises relative to the others, the effect is to encourage the others more than itself, thus restoring the balance.

Given the structure of a hypercycle, information can be replicated, and maintained selectively, in an amount greater than would be possible if the whole message was replicated as a unit. It is worth thinking a little more carefully about how selection will act on a hypercycle. Suppose there is a mutation converting SAVE to SAME. The maintenance of the message requires that SAVE be replicated more efficiently than SAME, or than any other mutant. If this is so, and if the same is true of all the other words, then selection will maintain the cycle. But can a hypercycle evolve? Remember that, in biological reality, all that matters is that the message should replicate: that is its only meaning. If GOD SAME THE QUEEN replicates faster than the original message, then

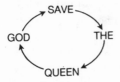

Figure 18. A hypercycle.
The words represent molecules. The rate of replication of each molecule depends on the concentration of the molecule immediately before it in the cycle. For example, the rate of replication of QUEEN increases as the concentration of THE increases.

it will become the prevalent type, and that will represent evolutionary progress. Consider, then, three types of mutation that might occur:

Type 1 is a mutation, say SAME, that replicates faster than SAVE, and also is better at stimulating the replication of THE.

Type 2, say SAFE, is also better at stimulating the replication of THE, but does not itself replicate any faster than SAVE.

Type 3, say SALE, replicates faster than SAVE, but is not so good at stimulating the replication of THE.

Of these three types, the first will be incorporated by natural selection, and this will improve the efficiency of the whole cycle. The second would not be incorporated, although it would improve the cycle if it were. Finally, mutants of the third type will be incorporated, and in so doing will tend to destroy the integrity of the cycle. Clearly, if natural selection is to improve the integrity of a hypercycle, and increase its complexity, we need to alter things so that mutants of types 1 and 2 are favoured, and type 3 eliminated.

Before discussing how this might have come about, I want to digress to compare a molecular hypercycle with an ecosystem (that is, a community of interacting plants and animals in the wild). Indeed, there is a sense in which a molecular hypercycle is an ecosystem, and an ecosystem is a hypercycle. Thus trees and earthworms form a hypercycle of two units: trees, by dropping their leaves, encourage the replication of earthworms, and earthworms, by digesting leaves and excreting nitrates, encourage the growth of trees. A type 2 mutant in a tree would be one that caused it to drop leaves more easily digestible by earthworms, but which did not directly help the tree. Would such a mutant spread in the population of trees? It would only spread if the additional earthworms produced helped the particular mutant tree more than it helped other trees. In this particular example that could be true, because leaves fall below the tree that grew them, and earthworms do not dash about. In contrast, a type 2 mutant in an earthworm, that helped trees rather than the particular earthworm, would not be likely to spread in the earthworm population, because the additional tree growth would help, not only the mutant earthworm and its offspring, but many other earthworms as well. It is for this kind of reason that ecologists are wary about drawing analogies between organisms, with their mutually supporting parts, and ecosystems: the organisms within an ecosystem exploit one another, but do not necessarily support one another.

The ecological analogy suggests that we expect mutually supportive interactions to evolve only when the interacting organisms do not move about too much, because only then will type 2 mutants be favoured by natural selection. This gives us the clue to the circumstances that would have favoured the evolution of more stable hypercycles on the primitive earth. Thus suppose that the reacting molecules were enclosed within coacervate droplets of the kind studied by Oparin, and that the growth and splitting of the droplets depended on the synthesis of the various components of the cycle. This is not unreasonable. In Oparin's experiments, particular kinds of molecule are concentrated by natural means within the droplets, and splitting is speeded up by synthetic processes—for example, the synthesis of starch. It would then follow that a droplet in which a mutant of type 2 occurred would grow faster, and one in which a type 3 mutant occurred would grow more slowly. By enclosing the reacting molecules within a protocell, we have ensured that selection acts on the whole structure, and not only on its components.

I must now try to summarize the argument so far. The purely chemical problems, of the origins of organic compounds and polymers, are well on the way to solution, following the experiments of Miller in the 1950s. The problem of hereditary replication is more difficult. The replication of nucleic acids in the absence of informed enzymes would happen, but with low accuracy. This raises the fundamental difficulty that, with high error rates, only very short molecules, probably less than 100 bases, could be maintained by selection, whereas much longer molecules are needed to code for the enzymes that could increase the accuracy of replication. A possible way round this difficulty is that a number of different replicating molecules might interact to form a hypercycle. Such a hypercycle would evolve further only if it was contained within some kind of bag. The experiments of Oparin and Fox show that such structures—coacervates and microspheres—could well have existed; that they could have concentrated within them the reacting molecules; and that their

The origin of life

growth and division could have been speeded up by the synthetic activity of the contained hypercycles. If so, natural selection could generate further complexity.

There remains one further difficulty, which may be the most severe of all: the relation between proteins and nucleic acids. Nucleic acids can replicate, but they are enzymically inert: chemically, they do not do anything interesting. Proteins are marvellous catalysts, but cannot replicate. Life depends on combining these two functions. Today the connection is made via the genetic code, and the process of directed protein synthesis. In this process the essential step, that in which an amino acid interacts with a nucleic acid, occurs when a particular amino acid is attached to a particular tRNA. The specificity of this attachment is achieved by an enzyme. This arrangement cannot be primitive. The first binding of amino acids to nucleic acids must have occurred in the absence of informed enzymes. At present we have little idea of how this could have happened: indeed, little is known of the chemistry of amino acid – nucleic acid interactions. We know that they are possible—for example we know that a protein 'repressor' will bind to a particular base sequence in DNA (see p. 66)—but that is about all.

The best guess we can make at present is that the primitive arrangement was that shown in Figure 19. This resembles Figure 4, on p. 17, which illustrated protein synthesis, but the primitive system differed in the following ways:

i) the messenger RNA molecule is also the molecule that is replicated: that is, it is also the gene.

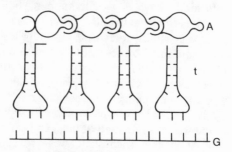

Figure 19. Primitive protein synthesis.

G represents a primitive gene, probably RNA; t represents primitive tRNA molecules, and A the amino acids to which they are attached.

ii) there is no ribosome to hold everything in position.

iii) there are no informed enzymes to attach the amino acids to the tRNA molecules.

iv) the number of kinds of tRNA molecules, and of amino acids incorporated into proteins, would have been much smaller than it is today.

I now turn to my last topic, the origin of the genetic code. One feature of the genetic code that must have been there from the beginning is that it is a triplet code. The reason for stating this so dogmatically is that it is a feature that could not alter once it was acquired. Further, as I argued on p. 19, once a triplet, or codon, had acquired a meaning—that is, once a tRNA molecule possessing the corresponding anticodon, and capable of being linked to a particular amino acid, had evolved—that meaning would not alter. (The meaning could be narrowed down, in the sense that a triplet might initially have coded for any one of several chemically similar amino acids, and later for only one of these.) That this is so is confirmed by the universality of the code today.

It is possible to make some rather more precise guesses. The first genes were probably rich in the bases G and C, and poor in A and U. This is because, as we have seen, non-enzymic replication is more accurate for G – C pairings. A second point emerges from a consideration of how the message may have been read. If you look at Figure 19, it will be apparent that (provided you do not mind having a few bases left over at the beginning and end) there are three different ways of reading the message, depending on how the 'reading frame' is chosen. If one of these ways produces a protein that has been favoured by natural selection, then the other two ways are likely to code for proteins with no useful function. In existing organisms, the choice of the correct frame is achieved as follows. Reading starts at a particular point, determined by a special sequence of bases, and the appropriate triplets are then chosen by counting off in threes. We know this because, if a mutation inserts a single extra base into a gene, then beyond that point the reading frame is shifted for every triplet, and every amino acid in the resulting protein is wrong.

Although this process of counting off in threes works today, it is hard to believe that it could be primitive. It seems more reasonable to suppose that the tRNA molecules arrived in any order, and attached to the message at any point they could. If so, a correct reading frame can still be achieved. Suppose that only some of the 64 triplets have a tRNA with a corresponding anti-codon. It is then possible to choose the 'meaningful' triplets in such a way that, no matter in what order they were arranged, none of the 'out-of-frame' triplets would be meaningful. The point will be clearer if I give an example. Suppose that GGC, GCC, GAC, and GUC were the only meaningful triplets. Then a gene might be, for example, GGCGACGACGCCGCCGGC. The first two out-of-frame triplets are GCG and CGA, which are not meaningful. Nor are any of the other out-of-frame triplets. Hence if the only tRNA's that exist have anticodons comple-mentary to the four codons listed, there is only one way in which the message can possibly be read. The tRNA's can arrive in any order, and no counting off is needed.

The use of these four triplets would also ensure a gene rich in G and C, as is required for non-enzymic replication. There are good theoretical reasons, then, why GGC and GCC may have been the first two triplets, and GAC and GUC the next two.[2] As it happens, there is an independent line of evidence leading to the same conclusion. In experiments of the kind performed by Miller, the two amino acids formed in the largest amounts are glycine and alanine, which today are coded for by GGC and GCC respectively; the next commonest are aspartic acid and valine, today coded for by GAC and GUC. It is hard to think that this could be a coincidence. Of course, this does not explain how the earliest decoding machinery worked in chemical terms, but it does suggest that our ideas about a G–C rich code which could not be read out of frame are along the right lines.

It is appropriate to have ended this book with a discussion of the code. I started by suggesting that there are two ways of viewing living things: as dissipative structures, and as entities capable of transmitting hereditary information. Proteins, and the chemical reactions they catalyse, are responsible for the flow of energy needed to maintain a dissipative structure. The precise base pairing of nucleic acids underlies hereditary transmission. These two aspects of life are brought together by the genetic code.

Notes

Chapter 2: Heredity

1. The historian R. C. Olby has recently suggested that Mendel was less clear about the nature of his factors, or genes, than I indicate here. He may be right, but I am inclined to give Mendel the benefit of the doubt.
2. In this sense, biology is more like history, in which accidents can be decisive—'because of a shoe, the horse was lost', and so on. However, it is easier to find general laws in biology. I doubt whether there is any principle in history which is at once as important and as nearly true as, for example, the central dogma of molecular biology, or the rules governing the behaviour of enzymes described in Chapter 6.

Chapter 3: Sex, recombination, and the levels of life

1. In fact, 'jumping genes' were discovered in maize over thirty years ago by Barbara McClintock, but the significance of her discovery was not widely appreciated until recently, when similar elements were found in prokaryotes.
2. This expectation may not be correct. There are a few cases in which it seems likely that a gene has been transferred from one eukaryote to another distantly related one. The most striking example is the presence, in leguminous plants, of a gene coding for a haemoglobin molecule, surprisingly similar to the protein which, in our own blood, combines reversibly with oxygen. However, it looks as if such distant gene transfer is so rare in eukaryotes that the distinction I have drawn between mice and bacteria is a valid one.
3. The analogy between the genes of a bacterium and the players in a football team should not be pressed too far. Probably the majority of genes in a bacterium are permanent components of that species— they are, if you like, the analogues of the home ground on which the team plays—and only a minority are temporary 'transfers'.
4. This paragraph makes several points. First, there have not always been two sexes, in the sense of individuals producing different kinds of gametes, eggs and sperm. Second, in so far as there was a first sex, it was the male sex, producing small motile gametes. Third, we have a satisfactory mathematical theory of the origin of sexual differentiation—that is, of the origin of females. There remains the question—

why only two sexes? Consider the case of three sexes, A, B, and C. Two situations can be conceived. First, each child would receive genes from all three parental sexes. This would mean that individuals would be triploid (that is, having three sets of chromosomes), and would require a process of gamete formation in which each gamete received exactly one of each kind of chromosome. It is hard to see how such a process could evolve, and equally hard to see what advantage it would confer if it did. The alternative is that any two of A, B, and C could produce a child. But if A, B, and C are differentiated, it would inevitably be the case that one pairing would be more efficient than the other two, and the third sex would be eliminated. There can, of course, be three 'sexes', in the sense of three kinds of individual, all of which are needed to produce offspring, but only two of which contribute genetic material: such a system is characteristic of the social insects (queens, drones, workers).

Chapter 4: The pattern of nature

1. If two species form hybrids on the rare occasions on which their ranges overlap (as do, for example, primroses and cowslips, or the common bladder campion and the sea campion, or, among animals, the crested and marbled newt), this merely serves to allow some exchange of genes between them. In plants, however, and much more rarely in animals, hybridization is sometimes accompanied by a doubling of the chromosome number, and this can give rise, in a single step, to a new species reproductively isolated from both parents.
2. Note that we cannot simply bring representatives of the two populations together in captivity and see whether they mate. Even if they did, this would not prove that they belonged to the same species, because, as in the case of the river ducks, species which remain distinct in the same area in the wild may hybridize in captivity.
3. It is interesting that, whereas vertebrates have settled on two pairs of limbs, insects have settled on three pairs. This is because they are descended from centipede-like ancestors with more than three pairs: the number of legs has been reduced to six, which is the smallest number that allows you to lift half your legs off the ground and not fall over.

Chapter 5: Problems of evolutionary biology

1. The main proponent of the idea that chance events in small populations are crucial in evolution is Sewall Wright, the only surviving member of the great triumvirate (with R. A. Fisher and J. B. S. Haldane) who founded population genetics. Because of the accident that Wright is American, whereas Fisher and Haldane were British,

evolutionary biologists in America often place great emphasis on chance events, whereas we British know better.

2. There is a bacterial plasmid (see p. 31) which carries two genes. One of these is active most of the time, and produces a protein which protects the bacterial cell against a particular kind of poison, known as a colicin. The second gene is switched on only when the bacterial culture is overcrowded. It then produces the colicin. The effect of this is to kill the bacterial cell, and the plasmids it contains: the plasmid has committed suicide. However, when the cell dies, colicins are released into the medium, and kill those bacteria which do not have the protective protein; that is, that do not carry the plasmid. Thus, by committing suicide, the plasmid ensures the survival of other plasmids genetically identical to itself.

Chapter 7: Behaviour

1. This conclusion is by no means obvious. It depends not only on the initial increase in the rate of turning in a bright light, but also on the fact that the animal 'adapts' to the light. However, the conclusion is correct.

2. Bees can communicate the direction and distance of a food source by dancing on a vertical surface. The angle between the direction of the dance and the vertical is equal to the angle between the direction of the food source and that of the sun, and is so interpreted by other bees.

3. A still harder question is whether only brains made of neurones can have feelings, or whether any other structure capable of performing the same operations would have feelings, whether it is made of neurones, or transistors, or whatever the brains of Martians are made of. If consciousness is an emergent property of matter organized in particular ways, it seems that any structure functionally equivalent to a brain should be conscious.

4. This experiment was performed by R. J. G. Morris. The interpretation I have given is his. However, on thinking about the matter further, I'm not sure it is justified, because there is a fairly simple set of rules that the rat may be following. Let us suppose that there are two recognizable objects, A and B, outside and above the tank, that are visible to the rat (if there are no such objects, then the problem is insoluble, with or without a cognitive map). Having once reached the platform, the rat can remember the angles of inclination, say α and β, of objects A and B. When released in future, it can obey the following rules: if angle of inclination of A is less than α, then swim towards A, and vice versa; if the angle of inclination of B is less than β, swim towards B, and vice versa; in combining these two instructions, pay more attention to the object with the greater angular

discrepancy. This would lead the rat to the platform. Of course, this does not prove that rats do not use a cognitive map, but it does show how hard it is to devise a decisive test. It may be that we shall need direct physiological evidence to prove the existence of a cognitive map in an animal. Some evidence of this kind does in fact exist.

5. Remember: The centipede was happy quite
    Until the toad in fun
    Said 'Pray, which leg comes after which?'
    Which brought its mind to such a pitch
    It lay distracted in a ditch
    Considering how to run.

Chapter 8: The brain and perception

1. The existence of single cells in the cortex that respond to particular features on the retina—for example, oriented lines—has suggested that there may be cells responsive to more complex patterns. The classic example is the imagined 'grandmother-detector'—a cell which would fire when seeing, or hearing, or being told about, your grandmother. Clearly, if such a cell exists, it must receive inputs from many sources: it cannot be in area 17. However bizarre the idea may seem, I think it will prove to be essentially correct. All computation depends on symbolic representation: the first thing you do when you write a computer program is to choose a set of symbols and decide what they will stand for. If thinking is at all like computing, then it too requires that we operate with symbols, which stand for the objects we are thinking about. The meaning of a symbol is then defined by the set of experiences that can activate it (for example, the sight or sound of your grandmother), and by the actions that the symbol can elicit (for example, saying the word 'grandmother'). It does not follow that the physical representation of the symbol has to be a single cell, or group of cells, but that does seem to be the obvious way of doing it.

2. The refusal has also been crucially important in the history of science. It was the refusal to accept that the movements of the planets are random that was decisive in the birth of modern science. Nevertheless, a willingness to treat some things as random can be necessary for scientific progress, as in the Maxwell–Boltzmann kinetic theory of gases, and in Darwin's theory of evolution. Unfortunately, there are no rules to tell a scientist which data to treat as meaningful and which as noise.

Chapter 9: Development

1. My colleague Helen Spurway discovered a remarkable genetic mutation in *Drosophila*, which she called 'grandchildless', because females carrying it had offspring which were normal in every way,

except that they lacked ovaries or testes, and hence had no children. It later turned out that the original grandchildless female lays eggs that do contain pole plasm, but in which the movement of nuclei into the pole plasm is delayed, so that no primordial germ cells are formed.

Chapter 10: The origin of life

1. There is a second possible source of free oxygen. High up in the atmosphere, water molecules are split by radiation, and the lighter hydrogen molecules escape into space, leaving free oxygen behind.
2. The proposed code, using only the codons GNC, where N stands for any one of the four bases, has another advantage. The complement of any message composed of these codons is also composed of the same codons. To see this, you must know a fact about RNA (and DNA) which I have not previously explained. RNA molecules have a 'polarity', in the same way that a poppet necklace has a polarity, with the balls pointing one way and the cups into which they fit pointing the other. When replication takes place, the new RNA chain has the opposite polarity to that of the old one. Consequently, allowing for polarity, the complement of, say, GUC is GAC, and not, as you might at first think, CAG.

Further reading

The books marked* were written for readers with little or no scientific training.

Chapters 1–5

* J. Cherfas, *Man Made Life*, Oxford: Blackwell, 1982
* J. Maynard Smith, *The Theory of Evolution*, 3rd edition, Harmondsworth: Penguin Books, 1975
  E. Mayr, *The Growth of Biological Thought*, Cambridge, Mass.: Harvard University Press, 1982
* I. Prigogine and I. Stengers, *Order out of Chaos*, New York: Bantam Books, 1984
* M. Ridley, *The Problems of Evolution*, Oxford: Oxford University Press, 1985

Chapter 6.

* J. Monod, *Chance and Necessity*, London: Collins, 1972

Chapters 7 and 8

* M. Boden, *Artificial Intelligence and Natural Man*, New York: Basic Books, 1977
  J. A. Fodor, 'The Mind–Body Problem', *Scientific American*, January 1981
  G. S. Frankel and D. L. Gunn, *The Orientation of Animals*, New York: Dover, 1961
* R. L. Gregory, *Eye and Brain*, London: Weidenfeld and Nicolson, 1966
* D. R. Hofstadter and D. C. Dennett, *The Mind's I*, New York: Basic Books, 1981
  D. Marr, *Vision*, San Francisco: Freeman, 1982
  B. F. Skinner, *About Behaviourism*, London: Cape, 1974
  N. Tinbergen, *The Study of Instinct*, Oxford: Oxford University Press, 1951

Chapter 9.

B. C. Goodwin, N. Holder and C. C. Wylie, *Development and Evolution*, Cambridge: Cambridge University Press, 1983

Chapter 10

* Francis Crick, *Life Itself*, London: Macdonald, 1982

M. Eigen, W. Gardiner, P. Schuster, and R. Winkler-Oswatitsch, 'The Origin of Genetic Information', *Scientific American*, April 1981.

Index